POWER, INSTITUTIONS,
AND LEADERSHIP
IN WAR AND PEACE

POWER, INSTITUTIONS, AND LEADERSHIP IN WAR AND PEACE

LESSONS FROM PERU AND ECUADOR, 1995–1998

DAVID R. MARES AND DAVID SCOTT PALMER

University of Texas Press   Austin

Requests for permission to reproduce material from this work should
be sent to:
    Permissions
    University of Texas Press
    P.O. Box 7819
    Austin, TX 78713-7819
    www.utexas.edu/utpress/about/bpermission.html

The paper used in this book meets the minimum requirements of
ANSI/NISO Z39.48-1992 (R1997) (Permanence of Paper). ∞

Library of Congress Cataloging-in-Publication Data

Mares, David R.
Power, institutions, and leadership in war and peace : lessons from
Peru and Ecuador, 1995–1998 / David R. Mares and David Scott Palmer.
    1st ed.
    p.   cm.
Includes bibliographical references and index.
ISBN 978-0-292-73569-9 (cloth : alk. paper)
ISBN 978-0-292-73570-5 (e-book)
    1. Politics and war—Case studies.   2. Political leadership—Case
studies.   3. Boundary disputes—Case studies.   4. Politics and war—
Peru.   5. Political leadership—Peru.   6. Politics and war—Ecuador.
7. Political leadership—Ecuador.   8. Peru—Boundaries—Ecuador.   9.
Ecuador—Boundaries—Peru.   I. Palmer, David Scott, 1937–   II. Title.

    JZ6385.M35   2012
    355.02—dc23   2011039279

CONTENTS

LIST OF MAPS AND TABLES

MAPS

TABLES

PREFACE

The origins of this volume go back to January 1995, when war broke out unexpectedly between Ecuador and Peru. Each of us had been following developments in both countries for some time and had spent extended periods in each conducting research and participating in conferences and programs, among other activities, during repeated visits. Over the course of our academic careers, we had written extensively on internal political and security issues in each country and in the wider region. One result of our long acquaintance with both countries, as well as our work on U.S. foreign policy issues, was that we already knew a number of the people who were or would become involved in the dispute as it slowly worked its way toward resolution. Working separately, starting within days of the outbreak of hostilities, each of us set out to try to understand what had happened and why.

After our field research led to several single-authored publications, we began discussing the possibility of a joint project, realizing that our individual work on the subject was complementary in multiple ways. Our first concerted work together occurred while Mares was on sabbatical at Harvard's Center for International Affairs and continued over the course of several years, with meetings at academic conferences and holing up for a few days at a time at each other's homes in Boston and San Diego to work through various parts of the manuscript. Almost inevitably, our efforts were interrupted at various junctures by other commitments. One of the unanticipated but positive outcomes of such delays, however, was to see how both parties' ability to forge a peaceful resolution to the conflict, after more than three years of difficult interactions, played out over the following decade.

As our work developed we found that the interplay of considerations of power, institutional relationships, and qualities of leadership among the multiple actors involved were the key factors that explained both the decision to go to war and the decision to make peace. Given the large number of ongoing land and sea boundary disputes that continue to involve most countries in the Western Hemisphere, we believe that our study provides insights into how such issues might be successfully addressed at the point when their particular dynamics have "ripened."

We gratefully acknowledge the assistance provided by the many individuals who met with us over the years we have been working on this project. They helped us in multiple ways to understand the complexities involved in working through the process of finding a solution to the hemisphere's longest standing boundary dispute. Many of these conversations also allowed us to appreciate the importance of both domestic and regional considerations that impinged on the dispute and on its eventual resolution.

While our debt to these individuals is acknowledged in the citations sprinkled throughout the chapters, we want to note our particular gratitude to Luigi Einaudi, U.S. representative to the guarantors under the Rio Protocol; to Ecuadorean diplomat and scholar Francisco Carrión Mena, participant in many of the discussions throughout the process and author of a detailed study of its denouement; to the Peruvian and Ecuadorean ambassadors to the United States, Ricardo Luna and Edgar Terán, who shared their insights and numerous official documents; to Gen. Franciso "Paco" Moncayo, who led Ecuadorean forces in the war, and to Gen. José Williams, commander of Peruvian Special Forces during the hostilities; and to former president of Ecuador Jamil Mahuad, who gave generously of his time to flesh out the fascinating details of "getting to yes" during the critical final weeks. Many others should be acknowledged as well, including military and diplomatic officials of Ecuador, Peru, and the United States, some named in the citations and others who requested anonymity. Each gave us additional perspectives and insights that added important details to our analysis.

Our acknowledgment of the contributions made to our final study by others would not be complete without recognizing the important role played by academic conferences and their organizers. Then University of Miami's North-South Center visiting scholar Tommie Sue Montgomery's 1997 invitation to academics and practitioners to discuss cases of peacemaking and peacekeeping in Latin America gave Palmer an early opportunity to present his research to that point on the Peru-Ecuador conflict and to receive valuable feedback. Another important gathering, at FLACSO (Facultad Latinoamericana de Ciencias Sociales)–Ecuador in Quito, this time related entirely to the Ecuador-Peru conflict, was organized by Dr. Adrián Bonilla in October 1998. We presented separate papers, as did several other leading Peruvian and Ecuadorean scholars, along with some key participants in the negotiations. With the unexpected announcement that the parties had reached a final peace settlement as our conference was concluding, the occasion took on a very emotional dimension.

Along the way, Palmer also benefited from a Fulbright Senior Research Professor award to visit Peru in 1998. This enabled him to observe at close hand developments there that threatened peace negotiations in their final stages and to conduct additional interviews. In addition, he would like to express his gratitude for invitations from Dr. Adrián Bonilla and Dr. Carlos de la Torre to teach month-long courses at FLACSO-Ecuador in 2006 and 2008. These extended stays in Quito provided opportunities to conduct various interviews to flesh out both civilian and military retrospectives on the conflict and its resolution.

David Mares would like to thank the Naval Postgraduate School at Monterey for an invitation to consider the implications for deterrence theory of the Ecuador-Peru war immediately after its termination. Adrián Bonilla facilitated three summers of fieldwork as a visiting scholar at FLACSO-Ecuador. Mares was also fortunate to offer a course on conflict and conflict resolution at the Diplomatic Academy of the Ecuadorean Ministry of Foreign Affairs in Quito; the feedback from these young diplomats, as well as from the officers at a symposium that Mares presented at the Instituto Nacional de Guerra, was fundamental in gaining perspective on Ecuador's position in the dispute. Andrés Mejía Acosta provided invaluable guidance through the public opinion data and thinking about the incentives facing Ecuadorean legislators. Jorge Domínguez's invitation to spend a sabbatical at Harvard's Center for International Affairs provided important time for the manuscript to germinate. At University of California, San Diego, the Academic Senate's Committee on Research provided funds for graduate student research support. Octavio Amorim Neto and Lydia Tiede provided excellent research assistance, and Jaime

Arredondo and Emilia Garcia came through with important last-minute editing.

At the end of our long journey, Theresa May of the University of Texas Press offered enthusiasm, guidance, and expertise for the launching of the book.

The book's cover design was inspired by longtime philatelist, scholar, and friend Jack Child (d. 2011) and his book, *Miniature Messages: The Semiotics and Politics of Latin American Postage Stamps* (Duke, 2008).

In spite of our best efforts, this study no doubt contains some errors of commission or omission, for which we take full responsibility. Whatever the limitations, we feel that our study provides an important perspective on conflict and conflict resolution and the forces involved. We would like to dedicate our book to those Ecuadoreans and Peruvians who have worked so hard to overcome the centuries of distrust and conflict which divided them and who were able, in the face of multiple challenges, to find a way to a full and peaceful resolution.

1 INTRODUCTION
Explaining Interstate Conflict and Boundary Disputes in Post–Cold War Latin America

In January 1995 fighting broke out between Ecuadorean and Peruvian military forces in a remote section of the Amazon that ultimately cost hundreds and perhaps even more than a thousand lives.[1] Ecuador refused to abandon outposts constructed in territory it disputed with Peru. As the fighting quickly escalated before becoming bogged down for some thirty-four days, first dismay and then determination gripped the Western Hemisphere. How could fighter bombers, helicopters, land mines, surface-to-air missiles, and thousands of troops be converging rapidly on a far-off section of the jungle? How could tanks, warships, and thousands more troops be mobilized in reserve, guarding sea lanes and potentially vulnerable points along a 2,000-kilometer border? Was there any way in which the international community could contribute to end the fighting and establish a lasting peace?

Some observers and policy analysts viewed this event as an anomaly in the "new" Latin America of dramatically lowered levels of overt ideological conflict, redemocratization, and economic integration.[2] After the U.S. "victory" in the Cold War and its undisputed influence in the Western Hemisphere in the 1990s, some were surprised that such violence could even be contemplated by Latin American states.[3] Other analysts interpret interstate violence as the result of some internal flaw in the political system: nondemocratic politics, "immature" democratic institutions, and populist leadership may contribute to diversionary war in times of political and economic stress.[4] These views do not explain the violence in 1995, however. Nor do they offer insights on how to minimize the likelihood of its future outbreak in a hemisphere still rife with disputed territorial and maritime boundaries, competition over natural resources, illegal flows of people and products across borders, and increasing arms purchases. One recent example is the military incursion by Colombia against a guerrilla camp on the territory of its neighbor Ecuador in 2008.

In the "new" democratic Latin America which emerged from the late 1970s through the early 1990s, sovereign jurisdiction disputes over land, territorial seas, and airspace continue to trouble the region (see the section "Disputes in the Western Hemisphere" below). In spite of presumed U.S. post–Cold War influence, in the 1990s military force was verbally threatened, displayed, and used at least fifty times in Latin America.[5] Nor should we believe that the few wars that have occurred in Latin America make it a uniquely peaceful region; a comparison of the number of wars after World War II puts the region in the middle of the pack.[6] In the specific case of Ecuador and Peru, the return and routinization of democratic practice were insufficient to overcome the long history of a festering border dispute between the two countries. Ecuador's political leaders and citizenry alike came to believe that war was a legitimate means to produce a "just" settlement. As fighting broke out in the midst of an electoral campaign for president in Peru, the leading opposition candidate, former United Nations (UN) secretary general Javier Pérez de Cuellar, advocated a much tougher response than the economic carrot and personal diplomacy approach that had been pursued up to that point by incumbent candidate Alberto Fujimori.

The use of large-scale military force over a sustained period clearly implies a decision by central government authorities. Not all disputes lead to violence, and not all violence escalates to war. In fact, most disputes imply neither war nor violence, even at low levels. In the case of the Peru-Ecuador border dispute, someone made a decision in the period from December 1994 to January 1995 that produced war and someone made another decision between August and October 1998 that brought peace. If we can explain who did what and why at both points, it may help us better understand the roles of power, institutions, and individuals in making war and peace and provide lessons that can be applied here and elsewhere in order to defuse a future threat more quickly.

The decision to make war or peace is influenced by both domestic and international factors. In each arena, power and institutions mediate interests and historical memories and shape the context in which choices will be made. Yet individuals are not mere captives of power and institutions—leadership matters. Its display can make a decisive difference in sorting out how options are articulated, perceived, and ultimately selected. Both the historical analysis of the politics of war and peace between Ecuador and Peru and the extended discussion of the relationship of this case to larger issues of conflict and its resolution illuminate the details of these interrelationships among power, institutions, and leadership in both domestic and international arenas.

This opening chapter has five sections. The first section discusses interstate disputes in the region. The data indicate that such disputes are neither rare in Latin America nor confined to the past. Thus thinking about how disputes become militarized, how they escalate to war, how the fighting stops, and how a dispute gets resolved has great relevance for the Western Hemisphere today. Subsequent sections examine our key variables: institutions, power, and leadership. The second and third sections analyze the argument that institutions are important determinants of international behavior, including war and peace. Domestic institutions are the subject of the second section, highlighting the importance of electoral incentives for politicians, bureaucratic politics, and military interests. The third section focuses on international institutions as facilitators of interstate cooperation. It points out the insufficiency of international institutions by documenting the widespread and elaborate institutional structure in the region at the outbreak of the war in 1995. The fourth section proposes ways to think about the relationship between power and institutions. We argue that power is relational and composed of a number of factors and thus is best understood in terms of a "strategic balance" among rivals. The fifth section discusses the importance of political leadership and proposes two characteristics that will determine to what degree leaders are constrained by the institutional context and power relations within which they operate. We develop hypotheses about the ability and willingness of political leaders to be innovative and take risks in dealing with international disputes.

Subsequent chapters deal with the core themes of external developments, domestic institutions, and individual leadership and statecraft. Chapter 2 presents the extensive historical background and details of the conflict itself. Chapter 3 discusses the domestic institutional elements in Peru and Ecuador that bear on our understanding of the dispute. Chapter 4 deals with the forces and factors that led to war. Chapter 5 considers the domestic bases for resolution of the conflict, while Chapter 6 offers a parallel analysis of the most relevant international elements. Chapter 7 discusses the lessons learned from the Ecuador-Peru dispute and how they might assist in dealing with and resolving the numerous ongoing disputes in the Americas.

In sum, this book presents a detailed analysis of the Ecuador-Peru border conflict and its resolution, but frames it within core themes from the literature on conflict resolution and draws lessons from this case for consideration in other ongoing disputes among nations.

Disputes in the Western Hemisphere

A significant number of interstate disputes continue to be present in the Western Hemisphere. These range from territorial and maritime disputes to the illegal flow of people and products across international boundaries. Some are quite active, others fester below a veneer of civility, and still others are consciously kept off the diplomatic table by all parties. The Ecuador-Peru experience, however, counsels us to avoid taking the non-violence or limited violence of these disputes for granted.

Of the thirty-five members of the Organization of American States (OAS), only three have not been involved in international boundary disputes (Bahamas, Jamaica, Paraguay), while thirty-two member nations (Antigua and Barbuda, Argentina, Barbados, Belize, Bolivia, Brazil, Canada, Chile, Colombia, Costa Rica, Cuba, Dominica, Dominican Republic, Ecuador, El Salvador, Granada, Guatemala, Guyana, Haiti, Honduras, Mexico, Nicaragua, Panama, Peru, Saint Kitts and Nevis, Saint Lucia, Saint Vincent and the Grenadines, Suriname, Trinidad and Tobago, the United States, Uruguay, and Venezuela) have had at least one such dispute during this period, in addition to disputes between French Guiana and Suriname and Great Britain and Argentina.[7] Counting disputes in Latin America has unfortunately been unsystematic and plagued by errors. For example, though Beth A. Simmons lists territorial disputes between Argentina and Chile as resolved in 1995,[8] they have not demarcated the Hielos Continentales (Southern Patagonian Ice Field). Chile issued a diplomatic protest in 2006 when Argentina produced a map with demarcations.[9] Nor has the El Salvador/Honduras dispute been resolved as she states: the 1992 International Court of Justice (ICJ) ruling on one aspect of the dispute ignored the Isla de Conejo, so the countries continue to have a boundary dispute.[10] Simmons lists the Bolivia/Chile dispute over Bolivia's outlet to the sea as settled in 1996, but that is clearly incorrect. The two countries have not even had ambassador-level relations since 1978. She also mistakenly claims that Argentina's dispute with Great Britain over the Malvinas/Falklands Islands was resolved in 1996,[11] but it is still ongoing as well.

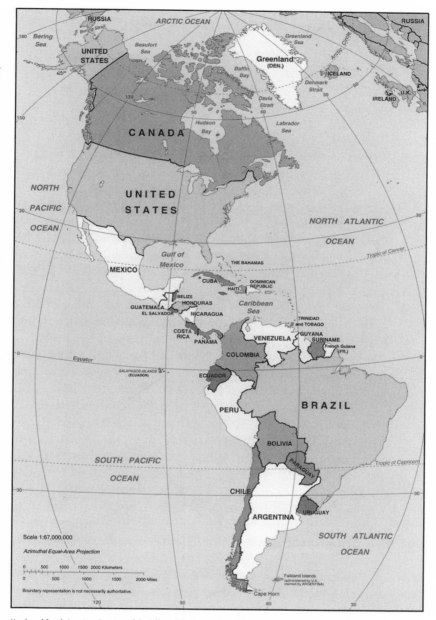

North and South America. Courtesy of the Office of the Geographer, U.S. Department of State.

South America. Courtesy of the Office of the Geographer, U.S. Department of State.

Although a number of dependencies of European states are found in the Caribbean Sea, only in South America are they involved in boundary disputes (French Guiana with Suriname and Great Britain with Argentina over South Atlantic islands, including the Malvinas/Falklands). From the list we can see that 91.4 percent (thirty-two of thirty-five) of the independent nations in the Western Hemisphere continue to have boundary disagreements of some type. Clearly, such disputes span the hemisphere.

Table 1.1. Boundary Settlements in the Western Hemisphere, 2000–2011

Countries and Year	Issue
Costa Rica/Colombia 2000	settlement of maritime boundaries in Pacific Ocean
Peru/Chile 2000	final implementation of 1929 agreement that provided Peru with a port in territory seized by Chile in the War of the Pacific (1879–1883)
Honduras/El Salvador 2003	reaffirmation of 1992 decision resolving land and maritime boundaries by ICJ
Trinidad and Tobago/Barbados 2006	maritime boundaries set by ICJ
Nicaragua/Honduras 2007	settlement of maritime boundary by ICJ
Venezuela/Trinidad and Tobago 2007	settlement of maritime boundaries
Nicaragua/Costa Rica 2009	judgment on San Juan River navigational rights by ICJ
Ecuador/Peru 2011	settlement of maritime boundaries by bilateral treaty

Source: "Peru Takes Possession of Chilean Port Terminal: Access to Arica Promised Since 1929."

Table 1.1 lists the boundary disputes that have been resolved within the last decade. Note that all eleven countries in Table 1.1 appear in the list of OAS member nations involved in disputes above, indicating that countries often have multiple boundary disputes. Disputes can be settled anywhere in the Western Hemisphere: three of the settlements came in Central America, one in the Caribbean, two in South America, one involved a Caribbean and South American country, and one involved a Central American and South American nation. Finally, by demonstrating that eight boundary disputes have been resolved in the first decade of the new millenium, Table 1.1 suggests that progress can be made in diminishing the number of outstanding disputes over time.

Tables 1.2 and 1.3 present the outstanding boundary disputes in the hemisphere. The disputes in Table 1.2 are actively contested. Notice the broad geographical distribution of these disputes: of the twelve that are active, six are in South America, three in Central America, and three in the Caribbean.

Table 1.3 lists disputes that are not currently active. Unfortunately, we lack the information to distinguish between disputes that are inactive because neither side wants to discuss the issue and those in which one side refuses to negotiate while the other is unwilling to resort to the unilateral use of force. Bolivia's desire for a sovereign outlet to the Pacific is an example of the latter. The United States is probably overly represented as a

Table 1.2. Active Contemporary Boundary Disputes in the Western Hemisphere

Countries	Disputed Issues
Argentina/Great Britain	Malvinas/Falklands, South Georgias, and South Sandwich Islands, including oil exploration in area
Bolivia/Chile	Bolivia seeks outlet to the sea; Río Lauca water rights
El Salvador/Honduras/Nicaragua	maritime boundary in the Golfo de Fonseca; ICJ advises that some tripartite resolution is required
Grenada/Trinidad and Tobago	maritime boundaries (currently being negotiated)
Guatemala/Belize	territory in Belize claimed by Guatemala; awaiting national referenda in both countries to submit to ICJ, but in OAS mediation
Guyana/Suriname	claims area in Guyana between New (Upper Courantyne) and Courantyne/Koetari (Kutari) Rivers (all headwaters of the Courantyne)
Guyana/Venezuela	entire area west of the Essequibo River claimed by Venezuela
Honduras/Cuba	maritime delimitation (treaty awaiting Honduran delimitation with Nicaragua)
Nicaragua/Colombia	waters around Archipelago de San Andrés y Providencia and Quitasueño Bank (at the ICJ)
Nicaragua/Costa Rica	San Juan River dredging
Peru/Chile	maritime delimitation (at the ICJ)
Venezuela/Colombia	maritime boundary dispute with Venezuela in the Gulf of Venezuela

Sources: CIA, The World Factbook 2010; U.S. Department of Defense, *Maritime Claims Reference Manual*, June 2008; International Boundary Research Unit; "Grenada, Trinidad Governments to Discuss Proposed Single Maritime Boundary," Caribbean Net News, March 15, 2010, http://www.caribbeannewsnow.com/caribnet/archivelist.php?arcyear=2010&arcmonth=3&arcday=15&pageaction=shownews (accessed April 12, 2010); "Foreign Affairs Minister to Table Belize-Guatemala Settlement Proposal to Cabinet," September 4, 2009, St. John's College, Belizean Studies Resource Center, http://www.sjc.edu.bz/belizeanstudies/newsmodule/view/section/6/id/8/src/@random4aa175eecb74c./ (accessed April 12, 2010); Nicky Pear and Alexandra Reed, "Dredging Up an Old Issue: An Analysis of the Long-standing Dispute between Costa Rica and Nicaragua over the San Juan River."

disputant, because it was easy to locate U.S. disputes. Other OAS countries undoubtedly have these sovereignty disputes with Western Hemisphere countries other than the United States, but they fail to show up in the sources consulted because they cover maritime and airspace disagreements which have not produced tensions. Consequently, we should take the thirty-one disputes listed in Table 1.3 as a minimum number. The inactivity of these disputes suggests a low level of primacy for the states involved. Any efforts to activate these disputes, even in order to try to resolve them, need to consider the risks that increasing the saliency of the issue might involve for the parties. Once again, with the exception of the United States, no region in the hemisphere stands out in this table.

As Table 1.3 indicates, the United States maintains boundary disagreements with twenty-two of the thirty-five members of the OAS. A number of these issues involve claims that foreign warships and planes must solicit permission to transit territorial seas and airspace. Although these types of disputes are unlikely to escalate to severe conflicts, it is U.S. military might that limits them, rather than something inherent in maritime and aerial sovereignty issues. In fact, the United States has actually engaged in provocative behavior in these cases by conducting "operational assertions" of its rights to go wherever it wishes without prior permission. Therefore we cannot assume that the response would be similarly peaceful if a Latin American nation were to engage in similar operational assertions. In the 1987 Caldas incident, for example, the presence of a Colombian warship in waters disputed by Venezuela produced military mobilizations and a two-week crisis.[12]

One of the interesting aspects of these boundary disputes, aside from their ubiquity, is that they can involve the closest of allies and be resolved even by the worst of enemies. To illustrate: the United States and Canada maintain numerous boundary disputes, while the United States and Cuba negotiated a treaty agreement over maritime boundaries in 1977. Although this treaty still has not been ratified, the parties have continuously exchanged diplomatic notes extending it.[13]

Migrations, contraband (including illicit drugs), guerrillas, and even disagreements about implementing treaties ratified by all sides can lead to severe conflict, including military action. Certainly, the U.S. invasion of Panama in 1989, in which over twenty thousand U.S. Marines captured Manuel Noriega at the cost of approximately five hundred lives in the name of democracy and the drug war, is a dramatic instance. But Latin Americans themselves have been willing to use force against their neighbors for reasons having little to do with territory or boundaries, as the 1937 massacre of twelve thousand to thirty thousand Haitian migrants by Dominican police forces demonstrates. In the 1969 war between Honduras and El Salvador that cost over three thousand lives, the border dispute was secondary to the migration issue. Venezuela and Colombia have seen their relations deteriorate in recent years over guerrilla, contraband, and migration issues rather than because of their boundary disagreements.[14]

Table 1.3. Currently Inactive Contemporary Boundary Disputes in the Western Hemisphere

Countries	Disputed Issue
Antarctic Treaty: Argentina, Australia, Chile, France, New Zealand, Norway, and the United Kingdom	defers claims, some overlapping, of parties to the treaty; the United States and most other nations do not recognize these territorial claims and have made no claims themselves (although the United States reserves the right to do so); no formal claims have been made in the sector between 90 degrees west and 150 degrees west
Brazil/Uruguay	Arroio Invernada (Arroyo de la Invernada) area of Rio Cuareim (Río Quarai) and islands at confluence with Uruguay River
Costa Rica/Colombia	Costa Rica waits to ratify Treaty for Caribbean delimitation pending Nicaraguan claims; has asked to join ICJ proceedings on Nicaragua/Colombia dispute
Guatemala/Honduras	delimitation of Río Montagua
Honduras/El Salvador/ Nicaragua	ICJ ruling on Golfo de Fonseca; advises that some tripartite resolution is required
Honduras/El Salvador	status of Isla de Conejo
Saint Vincent and the Grenadines/Trinidad and Tobago	maritime boundaries
Suriname/French Guiana	area between Rivière Litani and Rivière Marouini (both headwaters of the Lawa)
United States/Antigua and Barbuda	1982 Antigua and Barbuda Territorial Waters Act requiring prior permission for foreign warships to enter territorial sea/United States conducted "operational assertion" 1987
United States/Argentina	1961 limits of Río de la Plata claimed by Argentina as historic rights; 1967 Argentine territorial sea claim in excess of 12 nautical miles (nm) and not closing lines established for San Matias, Nuevo, and San Jorge Bays
United States/Barbados	1977 Barbadian Territorial Waters Act requiring foreign warships to obtain permission prior to transiting territorial sea/United States conducted "operational assertions" 1982, 1985, 1987
United States/Brazil	military exercises within the EEZ require consent of the coastal state; United States protested 1983 and 1988
United States/Canada	1906 Canadian claim of Hudson Bay as historic waters; 1967 straight baseline claims around Labrador and Newfoundland; 1969 straight baseline claims around Nova Scotia, Vancouver Island, Queen Charlotte Islands; 1986 straight baseline claims around Canadian Arctic Islands
United States/Chile	1985 Chilean claim of 350 nm from Easter Island and Sala y Gómez Island
United States/Colombia	1984 Colombian establishment of straight baselines/operational assertions 1988 and 1996

United States/Costa Rica	1978 Costa Rican law requiring foreign fishing vessels transiting EEZ to provide notification; 1988 Costa Rican establishment of straight baselines along its Pacific Ocean coastline; 1991 decree for foreign flag fishing to get permit before transiting Costa Rican waters
United States/Cuba	1977 Cuban decree establishing straight baselines/U.S. "operational assertions" 1985–1987; U.S. Naval Base at Guantanamo Bay
United States/Dominican Republic	1967 Dominican Republic straight baseline claims; claimed Escocesa and Santo Domingo Bays as historic bays; also enclosed Yuma, Andrés, Ocoa, and Ensenada de los Aguilas Bays; U.S. operational assertions 1987, 1991, 1992
United States/Ecuador	1966 Ecuadorean claim to 200 nm territorial sea/U.S. operational assertions 1979, 1980, 1985–1987, 1989–1994; 1971 Ecuador straight baseline claims; 1985 Ecuadorean extension of continental shelf to undersea Carnegie Mountain range, the Galapagos Archipelago
United States/Grenada	1978 Grenada requirement that foreign warships give notification before entering 12 nm territorial sea/U.S. operational assertion 1988; "certain aspects" of 1978 claim to 200 nm in EEZ
United States/Guyana	1977 Guyanese requirement that foreign warships give notification before entering 12 nm territorial sea/U.S. operational assertion 1988
United States/Haiti	1972 Haitian straight baseline claims/U.S. operational assertions 1986, 1987, 1991; 1977 Haitian claim to security jurisdiction in 24 nm contiguous zone; 1988 Haitian prohibition of entry "into ports, territorial waters and the EEZ if . . . transporting wastes, refuse, residues or any other material likely to endanger the health of the country's population and to pollute the marine, air and land environment"; Navassa Island; Haiti tried to activate prohibition of entry in 1998, United States rejected
United States/Honduras	straight baseline claims
United States/Mexico	1968 Mexican decree of straight baselines in Gulf of California; 1986 Mexican claim of northern Gulf of California as internal waters; 1999 exclusion of nuclear-powered and nuclear-armed ships from Mexican waters and ports
United States/Nicaragua	1979 act subjecting merchant ships transiting 200 nm territorial sea to internal laws of Nicaragua and international agreements; 1981 Nicaraguan requirement for clearance by aircraft and vessel transit of territorial sea; 1983 Nicaraguan decree of a 25 nm security zone requiring 15 days' notice for warships/military aircraft and 7 days' notice for civilian transit; United States conducted regular operational assertions from 1982 to 1993, 1998, and 1999
United States/Panama	1956 claim of Gulf of Panama as historic bay
United States/Peru	1947 Peruvian decree of 200 nm as territorial sea/United States operational assertions 1980, 1985–1988, 1990–1994
United States/Saint Vincent and the Grenadines	1983 Saint Vincent and the Grenadines act demanding prior permission for foreign warships to transit its 12 nm and the Grenadines territorial sea

United States/Uruguay	1961 limits of Río de la Plata claimed by Uruguay as historic rights; 1969 Uruguayan claim to 200 nm as territorial sea; 1982 restrictions on military exercises within EEZ
United States/Venezuela	1968 Venezuelan straight baselines claims; 1989 Venezuelan claim for 15 nm security jurisdiction in territorial sea; in 2000 Venezuela challenged U.S. vessels performing counterdrug ops in the Gulf of Venezuela; appears to indicate a de facto claimed historic bay status and/or enclosing straight baselines; U.S. protest against challenge of U.S. ships in the gulf in 2000
Venezuela/Dominica, Saint Kitts and Nevis, Saint Lucia, and Saint Vincent and the Grenadines	Exclusive Economic Zone (EEZ)/continental shelf extending over eastern Caribbean Sea

Sources: CIA, *The World Factbook 2010*; Department of Defense, *Maritime Claims Reference Manual*, June 2008; International Boundary Research Unit; Stephen Kangal, "T&T Must Expedite Maritime Border with Grenada," August 27, 2008, http://www.trinidadandtobagonews.com/blog/?p=587 (accessed April 12, 2010).

Domestic Institutions

Institutions are humanly devised constraints that shape human interaction by prescribing acceptable forms of behavior and proscribing unacceptable ones. These rules may be explicit and embodied in organizational structures or implicit and informal.[15] Institutions are more formalized in domestic politics because the structure of a polity reflects them. At the international level institutions are not as numerous; they may be more informal (as in the case of diplomatic conventions) and provide weaker incentives to guide international behavior.

Institutions are important factors in explaining both domestic and international behavior because political competition threatens to become violent whenever participants worry about their vulnerability to the misbehavior of other actors. At the domestic level such a situation may produce a collapse in domestic order, perhaps even leading to civil war. Internationally, the analog is interstate tension that can lead to war.

The danger of a breakdown in cooperation is diminished when institutions accomplish three tasks: (1) provide information concerning the likely behavior of others, (2) reduce transaction costs in collecting and evaluating that information, and (3) provide transparency in the actual behavior of others. The combination of these three elements enhances the credibility of the actors' commitment to cooperate, thereby reducing the danger of falling prey to partners who go back on their word.

The mere existence of institutions, however, does not automatically accomplish cooperation at either the domestic or the international level. (Clearly, democratic institutions have collapsed with depressing regularity in Latin America, and the United States has ignored the OAS on various issues of importance to Latin American governments.) The ability of institutions to have a significant impact is enhanced if three conditions are present. First, the relevant actors must expect to be around in the future to play the game again so that today's losses might be reasonably expected to even out over time. Second, the institutions must be able to create links across issue areas, thereby allowing actors to be compensated for losses in one area by gains in another. Third, the most powerful actors must support the institutions, even when they are constrained from pursuing their immediate short-run interests by the rules of those entities.

Democratic domestic institutions are by far the best for enhancing cooperation.[16] This type of institutional structure provides for greater accountability by leaders to a broader (though still not complete) spectrum of the citizenry and greater information and transparency through the concomitant presence of a free press. While most democracies do not function with total effectiveness, they can usually be expected to perform in a more satisfactory manner in these areas than their authoritarian counterparts.

Edward Mansfield and Jack Snyder argue that during the first decade of "new" democracies, two factors produce dramatically increased risks of war.[17] The first is that certain "imperfections" in the composition of political parties and functioning of the electoral system may allow democratic politics to become "hijacked" by special interests. These interests will seek private benefits and distribute their costs to the country at large. The second factor lies in the nature of these powerful special interests. If they are nationalists they are likely to use external threats to build broader support at home. Consequently, they are more likely to push foreign adventures that produce war.

This study looks closely at Ecuador and Peru to see if these democratic imperfections are present and if they played an important role on the road to war. That is not a self-evident claim. Ecuador's political system is not easily hijacked by special interests because the barriers for entry to congress and the presidency are low—parties can be easily organized and win high office. It is plausible to argue that Alberto Fujimori hijacked Peruvian politics in 1992–1993 with his *autogolpe* (self-coup) before the country returned to democratic forms in 1993–1994; however,

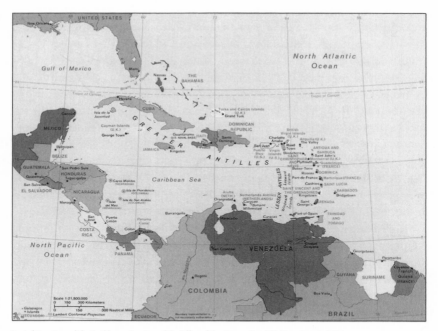

Central America and the Caribbean. Courtesy of the Office of the Geographer, U.S. Department of State.

his international strategy was not an aggressive military one. Rather than being a case of special interests hijacking politics, the imperfection that may matter most is the peculiarly Latin American tradition of "delegative democracy," in which leaders are given great leeway to undertake initiatives that promise, at least rhetorically, to solve the nation's most pressing problems.[18]

One of the lacunae in Mansfield's and Snyder's view of the behavioral implications of "new" democracies is that it provides no insights into why the same institutions that provide incentives for war can also produce peace and perhaps even a definitive resolution of the conflict. This omission would not be important if we could assume that the institutions of those who do not gain military victory will both change and move toward greater democracy as a result of war. However, there is no justification for making such an assumption. In the case at hand, both Ecuadorean and Peruvian democratic institutions began to weaken in 1997 after the Ecuadorean legislative coup against President Abdalá Bucaram and as President Alberto Fujimori started to restrict democratic practices and procedures in Peru in order to ensure continuance in power. Even in the context of weakened democratic institutions, however, the peace treaty was signed in 1998.

We argue that it is more fruitful to unpack the democratic institutional constraints themselves rather than focus on imperfections in their structure or on interest groups. Three factors in Latin American democracies help us understand how international behavior is affected by the institutional constraints of democratic politics: electoral incentives, bureaucratic politics, and military interests.

Electoral Incentives

In democratic polities, politicians face important constraints arising from the need to stand for periodic elections. The logic underlying this argument assumes that a politician's interest can be usefully condensed to winning elections. The claim is not that politicians have no other interests, but that they first need to be elected in order to accomplish whatever their goals are in politics. The politician thus needs to offer the voters what they want in order to be elected or reelected.[19]

To undertake a specific case analysis, we need a realistic model of electoral politics in that nation (see Table 1.4). This means examining the electoral system (regulations on running for office) and the party system, as well as public opinion on the relevant issue.

Term limits affect the electoral constraints on politicians in a democracy. Officeholders who wish to retain their position must be cognizant of voter preferences on policy issues. Presidents and members of congress who cannot run for reelection cannot be directly sanctioned via the ballot box. *Ceteris paribus*, presidents will face incentive structures that favor their pursuit of a personal policy agenda. The constraint on members of congress is still operative, however, in that they can seek election to other offices, whereas presidents rarely seek lower-level elective offices after their terms.[20]

Party strength should also affect electoral constraints on the executive. In political systems with strong parties, executives who cannot be reelected will feel constrained by the desire for their party's candidate to win next time. If the president is significantly constrained by the legislature, and party discipline is strong, even a president facing no reelection possibilities may feel constrained by voter preferences.

The strength of party systems varies across Latin America in terms of their level of institutionalization (stability in interparty competition, established roots in society, legitimacy among the populace, and stability of their organizational structure and rules).[21] In general, weak party

Table 1.4. Electoral Incentives of Leaders and Hypothesized Institutional Constraints on Leaders

Term Limits	Party Strength	Hypothesized Institutional Constraints
Yes	Weak	Very loose
Yes	Strong	Loose
No	Weak	Moderate
No	Strong	High

systems have negative implications for countries. Presidents are less likely to get support in the legislature because weak parties are not able to control congresses or lack the internal discipline necessary to sustain political coalitions to pass specific policy initiatives.

Such disadvantages for policy making may be mitigated to a degree in weak party systems because the presidents are not constrained by their need to support their parties in congress or to prepare them for the next presidential elections (see Table 1.4). Presidents in these circumstances can try to make policy on its intrinsic merits by appealing directly to the public for support. Going to the people, nevertheless, presents its own challenges, as demonstrated when Ecuadorean president Sixto Durán Ballén lost the referendum for economic reforms in 1995.

This discussion of domestic institutional context gives us a means to understand the role of history, a factor often simply alluded to as all-powerful until change actually happens. History and historical context may also play a significant role in given sets of circumstances. History is vast and complicated; what matters for politics is how history is interpreted and by whom. One way of thinking about how history matters is to associate different historical interpretations of an event with interest groups or the society at large. The weight of history on a particular issue will thus be influenced by the institutional constraints on the leader. The more highly constrained a leader is by his or her constituency, the more weight the corresponding interpretation of history carries.

The "weight of history" has implications for conflict resolution. The way in which history is interpreted will influence the nature of the side payments that are necessary for concessions on particularly tough issues. As this case demonstrates, the sense of insult and abandonment that Ecuadoreans and their military felt after the war of 1941 meant more than actual sovereignty over disputed territory. Understanding how that history was interpreted raised the possibility that resolution without actually modifying borders could be a viable option.

Bureaucratic Politics

The impact of bureaucracies on policy depends on how the decision-making unit is structured as well as the degree of the bureaucratic leadership's independence from the chief executive.[22] Although a foreign ministry might be assumed to have a significant impact on all foreign policies, policy making can be structured in such a way as to exclude or minimize the foreign ministry's influence. If the president selects ministers because of their professional qualifications rather than their personal ties to the chief executive, the bureaucracy will be more likely to influence policy.

The diplomatic corps in the United States and in many Latin American countries is very professional and capable. Yet a diplomat has very little independent bargaining power to use at home; the international reputation of a country is not a powerful tool in domestic politics. It is not in the interests of career diplomats in the foreign ministry, as agents of the president and subject to censure by the congress simply for policy disagreements, to blaze an independent path around new options to resolve issues as protracted and nettlesome as border disputes.

The diplomatic corps of a country, however, may play a fundamental role in proposing and developing options for other countries. Such an effort can help the national leadership of disputing nations to get something on the negotiating table that they themselves cannot offer directly. As the experience of the U.S. guarantor of the 1942 Rio Protocol, Luigi Einaudi, demonstrates in this case, innovative and skillful diplomats can have a major impact on helping rival nations find a mutually acceptable path to resolution.

Military Interests

The military is an agent of the government, but its firepower and reputation among particular segments of the population can render its subordination to government leaders problematic. Even well-established democracies find civilian control of the military to be incomplete.[23] Unfortunately, Latin America's history indicates that the issue of military subordination to democratic civilian control has not yet been settled. Thus military leaders have at least a de facto independent influence over policy making in areas they consider relevant to their role in national defense.

Because the military is a player on national security issues, we need to consider military interests when discussing interstate conflict in Latin

America. It is commonplace in the study of Latin American politics to perceive the military as xenophobic, antidemocratic, and focused on the short-term needs of its organization (arms, people, and pensions). But as the New Institutionalism paradigm demonstrates, the way in which self-interest is pursued varies by institutional context. In the current context of democratic politics, Latin American militaries are recognizing that professionalization and modernization of their force structure requires that they cede day-to-day operations of government to civilians and focus on their task of defending the nation. Democracy itself is thus not questioned, and the military's focus narrows to resources and nationalism.

What are the implications of the military's professionalization on the process through which border issues are negotiated? One might hypothesize that the military is unlikely to play the leading role in resolving a border dispute because of its intense nationalism, fears of territorial vulnerability, and concern over its budget and prerogatives. Two factors, however, suggest that the militaries may not be complete obstacles to peaceful resolutions of international disputes. Latin American militaries are not anxious for large-scale wars, because they can actually undermine the institution of the military. In war the outcome is often uncertain on the battlefield as well as at the peace table. Even military victory is no guarantee of "winning the peace." The economic costs of a major war to the country would likely be disastrous; with many basic needs of most Latin American populations unmet, governments might decide to cut the military budget and accept defeat. A comparison of the Argentine military's decision not to fight Chile in 1978 and its decision to fight Great Britain in 1982 shows that militaries can be very rational when it comes to war and still get it wrong, resulting in huge costs to the institution.[24] Consequently, in some conditions the military's professional interests turn it into a supporter of a comprehensive package resolving a border dispute.

If domestic institutions matter for interstate relations, should we be concerned about potential conflict in Latin America? The weakening of democratic practice in Latin America, while retaining democratic forms, is disappointing for many reasons. Paradoxically, however, the end of the Ecuador-Peru conflict as democracy was eroding in both countries is actually an encouraging sign. It suggests that peace and conflict resolution are possible even with very imperfect, even deteriorating domestic political institutions. To understand such an unexpected outcome more fully, we need to explore the role of international institutions in the process of peaceful settlement of disputes.

International Institutions

The institutional architecture of the Western Hemisphere is quite developed. The governments of Latin America, especially, have long viewed subregional, regional, and international bodies as important vehicles to accomplish their own foreign policy objectives while simultaneously serving to limit those of others. The Pan American Union (PAU, now the OAS) dates from the 1880s as a regional forum for working through various multilateral issues from the mundane (as in setting postage rates and procedures) to the significant (as in dealing with the principle of nonintervention). Ad hoc groups of states were often formed at the request of contending parties in the early to mid-twentieth century to help resolve disputes, from borders to wars. One such initiative was the establishment of a four-country guarantor mechanism (involving Brazil, Argentina, Chile, and the United States) within the Rio Protocol of 1942 that ended the 1941 Ecuador-Peru War.

The OAS succeeded the PAU in 1948 with a much broader mandate, including cultural exchanges, health programs, and dispute resolution among member states. While for many years it served more as a regional debating forum for official representatives, at moments of crisis foreign minister meetings sometimes assisted with the peaceful resolution of conflicts,[25] sanctions against members (for example, Cuba in 1962), or even authorizations for the use of force (as in the Dominican Republic in 1965). Too frequently, however, the OAS served more as an instrument of U.S. policy in the region than as a truly representative entity, often meriting Fidel Castro's reference to the OAS as the United States' "Ministry of Colonies."[26]

Nevertheless, due to a historic conjuncture of international, regional, and domestic factors in the 1980s, the OAS began to take on new and more autonomous responsibilities in Latin America. So, too, did other multilateral organizations, from the United Nations and new ad hoc groups of states with shared concerns to subregional free-trade areas.[27] The most important of these historic conjunctures included the following developments:

1. the winding down and end of the Cold War between 1985 and 1991,
2. the progressive democratization of Latin America beginning in 1978 and including fifteen countries by 1991,
3. the Iran-Contra scandal of 1986 in the midst of major U.S. unilateral involvement in Nicaragua,

4. Cuba's withdrawal of support for the government of Nicaragua and the rebels in El Salvador and Guatemala between 1988 and 1991, and

5. the election of an experienced internationalist (George Herbert Walker Bush) as president of the United States in November 1988, which produced significant new support for multilateral solutions to both international and internal conflict.

The overall result of such a felicitous combination of overlapping developments in the space of a few years was the strengthening of existing regional and subregional organizations and the creation of a number of new initiatives. With the Santiago Agreement of June 1991 (Resolution 1080), the OAS for the first time adopted a mechanism for consultation and response to internal threats to democracy in member countries. This mechanism was invoked, with considerable success, in response to *autogolpes* in Peru in 1992 and Guatemala in 1993 and to a civil-military standoff in Paraguay in 1994. Sometimes it was less successful, as in Haiti in 1991 after the military coup there. Resolution 1080 was strengthened in 2001 with the Democratic Charter, also known as the Declaration of Lima, which firmly established the protection of democracy in the hemisphere as a multilateral responsibility.

The Iran-Contra scandal, which involved illegal payments to the Nicaraguan rebels by White House officials at a time when the U.S. Congress had suspended government funding, both discredited the U.S. approach in Central America and served to immobilize the aggressive and unilateral policy. The controversy surrounding U.S. initiatives in Nicaragua since late 1981 had already produced an ad hoc regional response with the formation of the so-called Contadora Group, made up of Mexico, Panama, Colombia, and Venezuela, to serve as a moderating intermediary. Over time, four other South American countries that had recently returned to democracy—Peru, Argentina, Uruguay, and Brazil—complemented this ad hoc entity and came to be known as the Contadora Support Group.

In response to the perceived untoward meddling by both the United States and the Contadora Groups, the five Central American countries (for the first time in their history under elected governments at the same time) joined forces at the initiative of Costa Rican president Oscar Arias to produce their own peace plan in the Esquipulas Accords of 1986–1987 (also known as the San José Agreement or the Arias Peace Plan). In sum, multilateral regional groups made up of elected governments emerged out of the Central American conflicts to work together toward some definitive resolution that would end the violence. Their efforts were aided by changes in policies by key external actors—the blunting of U.S. policy due to Iran-Contra as well as the subsequent decision by Cuba to withdraw material support for the government of Nicaragua and the rebels in both El Salvador and Guatemala.

Under President George Herbert Walker Bush (1989–1993), U.S. policy shifted significantly toward support for multilateral initiatives to help restore peace in the region. Beyond further legitimization and reinforcement of OAS mechanisms to support democracy through resolutions, foreign minister meetings, and election observer missions, both the United States and its Latin American counterparts turned to the United Nations to secure support for peacekeeping initiatives in the region. While the UN had provided peacekeeping missions to other parts of the world for many years, it had never done so in the Western Hemisphere. With the assistance and direct involvement of the UN's first secretary general from Latin America, Javier Pérez de Cuellar of Peru, and of his successor, Boutros Boutros Ghali, this international body approved the establishment of four conflict resolution missions and one electoral observer mission to Central America and the Caribbean between 1989 and 1994. The UN Nicaragua 1990 Election Observation Mission was the first election oversight body under United Nations auspices. During the same period, the OAS also became much more involved, establishing twenty-four election observation missions between 1990 and 1994.

In each case, the role of international institutions has made a significant difference in the ability of the regional governments involved to establish and then maintain internal peace under inclusive democratic governments, even though problems remain, as in Guatemala and particularly in Haiti. Even with some difficulties, the era of multilateral responses to conflicts and electoral oversight undoubtedly was finally at hand. Such an array of domestic and international institutions can be expected to influence interstate behavior, but only in conjunction with the relative distribution of power among the rivals and the goals and skills of political leaders in specific situations, as illustrated by the failure to resolve the "coup" in Honduras in 2009.

The progressive development of such a web of international institutions should produce dramatically enhanced credibility of the signals being sent among rivals. This will undoubtedly be an asset when the parties are attempting to negotiate a solution. Yet we must also explain why the proliferation of such institutions in various parts of the region in the early 1990s did not prevent the Peru-Ecuador conflict from erupting into war in 1995 but did assist the parties over their almost four-year effort to achieve a definitive peaceful resolution of their dispute.

Power and Institutions

Institutions reflect power because the powerful can set the parameters for negotiating the creation of institutions. Once those institutions are developed, the powerful must find it in their interests to abide by those institutions if they are to influence behavior beyond what power itself can determine. Even in this situation, however, we cannot speak of institutions eliminating differences in power between competitors. Instead, as Peru's experience in this case demonstrates, institutions mediate power so that the powerful make concessions and pay a bit more for what they might well have gotten anyway.

If we think about institutions as setting contexts and providing incentives for certain behaviors and choices, then we have to think about power in terms of influencing what institutions might come into play in specific disputes as well as providing incentives for the institutions themselves to change the way they have defined an issue. Different forums may privilege different types of resources or choices. Ecuador tried to steer the dispute into institutions in which the issue could be discussed in terms of just settlements to war (for example, the UN), but Peru successfully kept the dispute in an institutional context defined by the Rio Protocol of 1942, even though this position ensured that the issue was not reopened for decades. Ecuador could not affect the way in which such a multilateral institution structured Peru's choices until it created a situation in which behavior (war) might escalate far beyond the abilities of the reigning international institution to manage.

Consequently, we need to think about power in terms of military, diplomatic, and economic resources. Power is relative, so we should think about how the distribution of these resources produces a "strategic balance" that determines how the costs and benefits of choices within that institutional context are distributed.

The *strategic balance* is a relative measure that includes but is not limited to the military balance. We use it here to refer to the factors which influence the likely costs produced by the strategies that each actor can use in particular disputes, rather than in its more narrow military sense (as in "strategic nuclear weapons"). As numerous studies of the conflict behavior of small states have demonstrated, a focus on the *absolute* capability of a nation, even incorporating nonmilitary factors, is inadequate for analyzing interstate conflict dynamics.[28]

The appropriateness of a measure of the strategic balance depends upon the particular political-military strategy being utilized and the political-military strategy being confronted. The strategic balance is defined by the resources that are relevant to those strategies and thus helps us understand the bargaining situation between the actors. While others have made this point by using variations in military strategy, risk assessments, and time frames,[29] we add diplomatic and economic factors to the range of relevant resources. Because of incomplete information, however, the strategic balance is never entirely clear to either party.

The *military balance* is a traditional concept for investigating power relations among states and includes the quality and quantity of personnel, the type and quantity of armaments, and doctrines for utilizing those resources. Studies of the foreign policy of great powers tend to emphasize the quantitative aspect of such resources because the social and economic disparities that underlie qualitative differences among great powers are not large. But the experiences of Iraq in the Gulf War, Israel in the Middle East, and Chile in South America demonstrate the importance of quality differentials where they exist.

Ultimately, military power might render institutions irrelevant by presenting all parties with a fait accompli. Institutions would simply reflect the terms of the course of the war. Yet military power may also be used to influence choices within a specific institutional context. If institutions matter, we need to think of military power not simply in terms of winning a war but also in terms of influencing the distribution of costs and benefits that accompany each policy choice.

The relevant *diplomatic resources* revolve around the ability to garner external support for, and blunt external criticism of, one's strategy in the dispute. This is affected not just by the skill of the diplomatic corps but also by the standing which one's position on the disputed issue has in the international political order of the era. Great powers may claim that their interests and values are universal, use force in the defense of those interests, and face little international sanction. Smaller powers, however, must couch the defense of their interests within the context set by the reigning political order of the great powers or be prepared to face international sanctions.

When smaller states can link their actions to the interests of great powers, new opportunities for advancing their interests arise. It may be possible to gain support for the use of force, aid in defending against a rival's use of force, or perhaps even increase international pressure on the rival to negotiate the previously nonnegotiable. Alternatively, when a small state has interests that are of minor consequence to the great powers, its rival's diplomacy might serve to convince the great powers that any benefits they might garner from becoming involved in the dispute would be outweighed by the associated costs.

Economic resources include both those that can be used in a non-military way to influence behavior by a rival and those for building up national capacity to use military force. When economic leverage is sufficient to gain one's goals at acceptable costs, force is unlikely to be used.[30] But when that economic leverage is deemed insufficient, the way in which economic resources affect a state's ability to mobilize, use, and resupply military forces becomes paramount.

A state's economic infrastructure (railroads, highways, and airports) can dramatically affect the logistical costs of using force. The ability to raise revenue for defense can be an important consideration, because it highlights the domestic opportunity costs involved in using force, thereby making it more likely that opposition to its use will form. For example, the inability of a state to tax the wealthy in a poor country imposes a severe constraint on state expenditures. Military expenditures thus come more openly at the cost of expenditures on economic and social welfare. When domestic elites are focused on moderating the polarization of society, they are unlikely to support a leader who wishes to spend the government's meager resources in militarized bargaining with a neighbor.

Honduras in the 1960s and 1970s provides a good example of this economic constraint on the use of force and helps explain why El Salvador believed it could quickly defeat Honduras with a blitz in 1969. Honduras also illustrates how diplomacy might overcome this constraint. After the 1979 victory by the Sandinistas in neighboring Nicaragua, the United States flooded Honduras with both economic and military aid. Economic constraints on the military capabilities of Honduras were thus overcome until the United States perceived that the Sandinistas were defeated and thus lost interest.[31]

While institutions and power influence choices, ultimately such choices are made by people, not by institutions or resources. So it is also necessary to consider the role of key leaders in dealing with conflict situations.

Political Leadership and Statecraft

Political leaders and diplomats respond to the incentives embodied in the political institutions within which they interact. As already noted, the major institutional factors facing presidents in Latin America will be those that influence elections, the existence of entrenched bureaucracies, and the military's willingness to abide by the particular political structures of authority reigning at the moment. Within these constraints, a leader's communication and organizational skills, as well as vision in creating opportunities for cooperation, can have a dramatic impact on conflict.[32]

In short, it is clear to us that political leaders, constrained as they may be, retain important discretion and an ability to make a fundamental difference. Those who do are often heralded as "political entrepreneurs." It is imperative for the analyst of international affairs to understand this factor in as systematic a way as possible.

The leadership qualities that should matter can be usefully summarized as follows:

1. The ability to innovate. Innovation is important because when war occurs it means that a dispute has festered: other strategies for resolution have been attempted, but failed. The absence of new ideas on how to think about or distribute the benefits at stake suggests a continuation of diplomatic failure and the possibility of violence to break the stalemate.

2. The willingness to take risks. Risk acceptance is important for two reasons. If a leader has new ideas but is timid about putting them on the agenda, the institutional constraints will determine whether or not he or she does so. Only under very loose constraints will a risk-averse leader propose new ideas that do not have the possibility of immediate support. In contrast, a leader who is risk acceptant and has new ideas will be quite willing to seek to create the political conditions that mitigate the institutional constraints *at least on this issue*.

Table 1.5 presents the hypotheses that are relevant for the Ecuadorean and Peruvian cases from 1995 to 1998. We determine whether leaders are innovative or risk acceptant by analyzing their behavior in prior crises on matters other than the border dispute. The contentious economic and political history of Latin America provides us with two excellent issues by which we can gauge the relevant personal characteristics of Peruvian and Ecuadorian presidents: the democratization process and economic reforms.

How do these incentive structures and personal characteristics affect policy making in the Ecuador-Peru case? With no reelection and a weak party system in Ecuador, we should expect innovative and risk-acceptant presidents to promote policies that they believe are correct and will strengthen their place in history, even at the expense of short-term

Table 1.5. Hypotheses Relating Leader Characteristics and Institutional Constraints

Risk Taker/ Innovative	Risk Taker/ Not Innovative	Innovator/ Risk Acceptant	Innovator/Risk Averse
Pushes new ideas while seeking to alter institutional constraints	Takes full advantage of institutional permissiveness to put issue on agenda but pushes old ideas	Pushes new ideas if institutional context is permissive	Reacts to events by sticking to traditional policy positions; strong hand of history

costs. Under these circumstances, presidents will be more susceptible to breaking with tradition if medium-term benefits are likely. But if the president is too cautious or lacks vision, the "strong hand of history" will make it unlikely that the obstacles to resolution will be overcome.

When reelection of a president is permitted, some of the incentives for a president change: the payoff from a risky policy would need to occur by the time of the next election for which the president is eligible (nonconsecutive terms in Peru until 1993, consecutive once until 2001; one nonconsecutive term in Ecuador after the 1998 reforms). Since the risk to the president is increased, the payoffs need to be proportionately greater than they were before. President Fujimori had incentives to settle the border dispute, given his desire to secure re-reelection on the basis of a dubious interpretation of the 1993 Constitution. An Ecuadorean president after 1998 who was innovative and risk acceptant could push for a historic solution if the payoffs were expected to materialize by the time the president could run again.

Conclusion

It is our perception that the war, subsequent peace, and ultimate resolution of the conflict between Ecuador and Peru are best understood as a rational process in which both domestic and international factors played fundamental roles at every step. We did not find that entrenched bureaucracies constituted a major influence on policy making, and our previous work indicates that the military was not a major autonomous factor in decision making on the border after the return to democracy. President Fujimori had few institutional constraints after his reelection in April 1995, with a weak party system, a largely re-created bureaucracy in the aftermath of hyperinflation, and a subservient military weakened by civil war, presidential favoritism, and massive retirements due to earlier economic distress. Ecuador's heads of state were very much limited in their ability to act by a multiparty congress, a diffuse bureaucracy, and a strong military, so they needed to consult regularly with these key actors to ensure consensus.

Among the critical factors that we have identified are the multiple developments beyond either country's borders in the late 1980s and early 1990s. In combination, they generated a dramatic increase in both the number of new multilateral organizations operating in the Western Hemisphere and the scope of activity and responsibility of existing multilateral bodies. While many important foreign policy relationships remained within more traditional bilateral frameworks, the multilateral arena offered new opportunities and alternatives for individual states (including Ecuador and Peru) as well as an emerging new diplomatic ethos further legitimating interaction among states to accomplish important foreign policy objectives.

Given the significant number of internal and external disputes, it is understandable that individual governments (and contenders for power as well) would find that multilateral mechanisms would be legitimate instruments to help them resolve these issues. In their moment of crisis, even as the 1995 war was breaking out, both Ecuador and Peru turned to the long-standing but also long-eschewed multilateral instrument of the guarantor mechanism within the Rio Protocol to try to find a solution to the dispute that had eluded them for decades, even centuries.

Another key factor that we have identified for conflict resolution is the operation and effectiveness of domestic institutions. At the moment the outbreak of hostilities between Ecuador and Peru occurred, both countries had functioning if imperfect democracies. Ecuador's had been operating since the transition from military to elected civilian rule in 1979. Peru had been democratic since 1980 with a brief but significant interruption in 1992 due to President Fujimori's *autogolpe*; by the end of 1993 it had been put back on track, with a new constitution and democratic procedures after intense OAS pressure through Resolution 1080.

In both countries, the chief executive took on the responsibility of trying to find an outcome that would satisfy national objectives. In Ecuador, with power more fragmented and dispersed, this meant constant consultation by the executive with key domestic actors, including congress, the military, business, and the media. In Peru, in contrast, with power concentrated in the presidency and with a majority in congress, consultation took place within a very restricted circle that included only a very small number of key officials in the executive branch. In a dramatic display of delegative democracy at a critical moment, both heads of state took the initiative to break through a final and apparently insurmountable impasse, as we shall see.

A third key factor is the exercise of individual leadership and statecraft within the inevitable constraints of domestic and international institutions. In the denouement of the Peru-Ecuador dispute and its resolution, Ecuador's President Durán Ballén made a critical decision at the outset of hostilities in January 1995 by once again accepting for his country (after a 35-year hiatus) the Rio Protocol as the instrument to serve as the foundation for negotiation. His successor, President Bucaram, committed Ecuador to a definitive resolution, gave up Ecuador's historic claim to "sovereign" access to the Amazon, and made the first state visit to Peru in his country's history. Jamil Mahuad, who succeeded Bucaram as president, took a bold personal initiative and accepted a final resolution that gave his country only symbolic access to hitherto disputed territory.

For his part, President Fujimori overcame great internal resistance to any negotiation with Ecuador at all by admitting that there was a problem in the first place. In addition, he met with the successive heads of state of his northern neighbor to bring a personal dimension to the process. At a crucial moment, he thwarted his military commander in chief's plan to reinitiate hostilities unilaterally at the border. With the process at its final impasse, he sacrificed his foreign minister and accepted the symbolic territorial concession that permitted a definitive resolution.

Within the multilateral mechanism of the guarantors, U.S. representative ambassador Luigi Einaudi served as the key actor. He devoted his full time and energies to coordination among his counterparts from Brazil, Argentina, and Chile, finding imaginative solutions to various impasses and serving as a constant and constructive presence even when there seemed to be no possible way to reconcile the differences between the parties.

All parties were constrained by deep-seated historical memories, physical limitations, the uncertain realities within their own countries, and the procedures imposed by international treaty. At critical junctures, however, each of these key figures exercised leadership that made important, even vital contributions to an eventual settlement.

These three core themes are developed in the chapters that follow. We turn first to a consideration of the background and details of the conflict itself, then consider the domestic institutional elements in each country, the factors leading to war, and the domestic bases for conflict resolution. We end with a parallel analysis of the relevant international elements.

2 TWO NATIONS IN CONFLICT

Like any pair of countries, particularly two that share a common border, Ecuador and Peru have had both converging and conflicting interests throughout their historical relationship. In this chapter we situate one particular conflict, the border war of 1995, in its historical context to help understand why the conflict characteristics loomed so large in this relationship. That context has three important components: one internal to each country, one bilateral, and a third regional in nature. With the 1995 war placed within these broader historical dynamics, the second part of this chapter explains why this war broke out between two elected governments and the elements that contributed to a final resolution. It is important to provide such a description of events over the decades to lay out the larger context that enables us to analyze the most important factors that ultimately combined to make possible a definitive resolution of the dispute.

Historical Context

The origins of this border dispute go back to the Spanish colonial administrative boundary designations, first between *audiencias* of the Viceroyalty of Peru and then (under the Bourbon Reforms of the 1700s) between this established viceroyalty and the new Viceroyalty of Nueva Granada carved out of its northern territories. At the time, administrative bound-aries co-existed with distinct ecclesiastical and military jurisdictions without creating major problems for the Crown. With few exceptions, the borders of Spanish colonial districts passed through sparsely inhabited and/or jungle areas largely beyond the effective reach of authorities lo-cated in the capital cities. For decades, if not centuries, the exact location of these borders was not of great import—they were all within the Span-ish colonial domain, after all, and the areas had no significant resources of value to the Crown to be extracted. As a result, precise boundary delin-eation appeared unnecessary and was not undertaken.

Independence created a new context in which borders suddenly mat-tered very much. During the wars of liberation, some areas ostensibly controlled by Quito joined the Peruvian armies, rather than those fighting farther north.[1] Peruvian authorities claimed that such acts represented self-determination to constitute a part of Peru. After independence, Peru attempted to seize further areas, including the major port city of Guaya-quil, but was defeated at Tarqui in 1829 by the forces of Gran Colombia (which at the time included Colombia, Venezuela, and Ecuador). Peru re-nounced some territorial claims in the Pedemonte-Mosquera Protocol of 1830, but the Congress of Gran Colombia did not ratify it. Gran Colom-bia broke into three countries shortly thereafter, and Ecuador sought to make the protocol effective on its own. However, Peru rejected Ecuador's initiative and its claim to the boundaries of a now defunct state. The dis-pute was papered over by the 1832 Pando-Noboa Treaty, with both sides interpreting the phrase "present limits" according to their own interests.

The area in dispute was remote, with no infrastructure and large-ly inhabited by indigenous peoples, yet it was perceived to have great economic potential. The Amazon River offered access to the potentially rich Amazon basin and a trade route to the Atlantic for South American countries on the Pacific coast. In 1860 Ecuador attempted to compensate European creditors with land in the Amazon. Peru attacked, forcing one defeated leader to recognize Peruvian claims, but other Ecuadorean lead-ers repudiated the 1860 Treaty of Mapasingue. Interestingly, although Chile sought Ecuadorean assistance during its two wars with Peru (1837 and 1879), Ecuador maintained strict neutrality.[2] Chile won both wars, so Ecuador may have lost an important opportunity to resolve the territo-rial issue in its favor.

Ecuador attempted to cede Amazonian land for debts again in 1887. This time Peru (recently defeated by Chile in the War of the Pacific, 1879–1883) and Ecuador agreed to negotiate their differences; unresolved issues were submitted to the king of Spain for binding arbitration. In 1890 the Peruvian executive granted Ecuador access to the Marañón River and the Ecuadorean congress quickly ratified the García-Herrera Treaty, but Peru's congress demanded a renegotiation. Between 1900 and 1904 a series of military clashes occurred in the region due to the expansion of rubber and gold exploitation and Peru's increasing integration of the region into the national economy. Diplomatic relations between Ecuador and Peru were severed for a time.[3] In 1905 Ecuador signed a secret treaty with another of Peru's territorial rivals, Colombia, in which each guaranteed the other's territorial integrity.[4] There was also an informal understanding in both Ecuador and Peru that Chile might help Ecuador in any conflict with Peru.[5]

In an effort to avoid war, the parties again turned to the king of Spain. In 1910 rumors that the king's advisors found Peruvian legal arguments compelling produced riots in Ecuador.[6] President Gen. Eloy Alfaro rejected the arbitration, called for new bilateral negotiations, and declared Ecuador's willingness to fight to preserve its Amazonian character. Both countries mobilized troops, and an arms buildup ensued. Argentina, Brazil, and the United States mediated, suggesting that the dispute be taken to the Permanent Court of Arbitration at The Hague. Peru accepted, but Ecuador called for direct negotiations.[7]

Although the king of Spain did not render a judgment, the basic outline of his view now defined the issue. Any juridical examination would most likely reproduce the king's view. Peru adopted arbitration as a fallback position if direct negotiations failed, while Ecuador sought to avoid juridical settlement. Of course, Peru would have no incentive to concede anything in bilateral negotiations, preferring the status quo (in which it occupied large sections of the disputed territory). Ecuador kept the level of tension on the border high in hopes that other Latin American states and the United States would insist that Peru accommodate Ecuador (in the twenty years before 1910 only three militarized disputes occurred; during the next eight years, they were constant).[8]

Despairing, Ecuador modified its strategy in 1916: it settled a dispute with Colombia in the Amazon to solidify a potential alliance against Peru. New attention centered on the United States in the wake of World War I. Both Ecuador and Peru hoped that the United States could obtain a "just" settlement in territorial conflicts in which they were the weaker party (meaning a settlement that would give the weaker party a more satisfactory outcome than could be garnered in bilateral negotiations between Ecuador and Peru and Peru and Chile). Because the United States consistently advocated peaceful resolution of conflict (even to the point of using its own military might to impose peace on Central America and the Caribbean),[9] Peru and Ecuador avoided militarized disputes through the 1920s.

The promise of diplomacy proved ephemeral. Peru, facing a new war scare with its powerful nemesis Chile, enticed Colombia to abandon Ecuador with a better deal in 1922: sovereignty in the disputed area north of the Amazon river in exchange for the territory previously ceded by Ecuador to Colombia. As a result, Ecuador severed relations with Colombia.[10]

In 1932, however, Peru escalated a minor border incident in the Leticia region into a major conflict with Colombia.[11] Colombia's military success encouraged Ecuadorean diplomatic and military posturing. As an "Amazonian" nation, Ecuador tried inserting itself into the Leticia cease-fire negotiations, but Peru blocked it. Military confrontations between Ecuador and Peru revived in 1932. Ecuador and Peru resumed diplomatic negotiations in 1933 and appealed to Washington for assistance, beginning in 1936, but increased border clashes in 1938 ended negotiations.

Ecuador's internal political situation continued to be unstable in the 1930s, and its military languished in domestic political struggles. In contrast, Peru began to climb out of the era of dictatorship. Peruvian officers blamed the politicization of their institution during the days of authoritarianism for the Colombian defeat. They resolved to professionalize themselves for their proper mission: the defense of national territory. This asymmetry in political development would have dramatic consequences.[12]

The 1941 War, the Rio Protocol, and Stalemate

In the late 1930s the Peruvian military saw an opportunity to resolve a territorial issue and end a history of defeat in regional conflicts. Ecuador was cognizant of its own relative weakness and responded by establishing small frontier outposts in the disputed territory to serve as tripwires, hoping to trigger international intervention.

In 1941 Peruvian troops, tanks, and planes swept across the disputed regions, penetrating deeply into Ecuador itself.[13] Ecuador appealed for diplomatic aid but encountered a United States and Latin America preoccupied with the wars in Europe and the Pacific. Peru threatened to occupy the entire territory it had seized until Ecuador recognized Peruvian claims in the Amazon. Bowing to pressure for inter-American solidarity, in January 1942 Ecuador accepted the Protocol of Peace, Friendship, and Boundaries of Rio de Janeiro, signed by both parties at the Third Consultative Meeting of the Ministers of Foreign Affairs of the American Republics in Brazil and ratified by their congresses the following month.[14] Argentina, Brazil, Chile, and the United States were designated as "guarantors" of the treaty, which denied Ecuador sovereign access to the Amazon River.[15]

The subsequent territorial adjustments of the Rio Protocol of 1942 were very modest from Peru's perspective because they were quite close to those accepted by both sides as the status quo during the previous diplomatic negotiations (see the historical time line at the end of this chapter).[16] From an Ecuadorean perspective, however, the status quo line of 1936 merely recognized that Peruvians had penetrated deeper into disputed territory, not the legality of that de facto possession. Thus Ecuador had proposed in 1937 that "[i]f Peru refused arbitration by President Roosevelt, an agreement be made on the line of the García-Herrera Treaty [of 1890]."[17] This treaty, negotiated between the two executives and ratified by the Ecuadorean but not the Peruvian congress, would have given Ecuador sovereign access to the Marañón and significantly more territory along the Napo River.

The main reason the dispute had not been resolved is that the parties took very divergent positions, which had less to do with any specific strategic or economic value associated with the areas in question than with a strong streak of nationalism. Ecuador believed that it continued to have sovereign rights to an area that Peru considered to be part of its own national territory. From the Peruvian perspective, any such rights were definitively terminated after the 1941 war by the January 1942 treaty. Peru also held that Ecuador's rights were further circumscribed by the 1945 Captain Braz Dias de Aguiar arbitral award accepted by both parties.[18] Ecuador based its claim on a combination of historical precedents (most particularly, the concept of *uti possidetis*),[19] treaty signing under duress, and treaty inapplicability due to geographical anomalies (see below). Most specialists considered Ecuador's position to be the weaker one under international law, because the 1942 protocol is a binding treaty under international law and postdates the early nineteenth century application of *uti possidetis* to the frontiers of the newly independent states of Spanish America.

The Rio Protocol, written and signed as World War II was beginning to reach the shores of the Western Hemisphere, provided for guarantor responsibilities that appeared at the time to be rather modest—to help the parties to the conflict get a boundary laid out and marked and if necessary to assist in addressing any disputes which might arise. The wording of the 1942 treaty is quite specific on this matter: "[A]ctivity of the United States, Argentina, Brazil, and Chile shall continue until the definitive demarcation of frontiers between Peru and Ecuador has been completed" (article 5). It also specifies navigation rights for Ecuador, "on the Amazon and its northern tributaries, the same concessions which Brazil and Colombia enjoy, in addition to those which may be agreed upon in a Treaty of Commerce and Navigation" (article 6). Furthermore, "any doubt or disagreement which may arise in the execution of this protocol shall be settled by the parties concerned with the assistance of the representatives of the United States, Argentina, Brazil, and Chile in the shortest possible time" (article 7). It adds that "the parties may, however, when the line is being laid out on the ground, grant such reciprocal concessions as they may consider advisable in order to adjust the aforesaid line to geographical realities" (article 9). The document goes on to state that "these rectifications shall be made with the collaboration of the representatives of the United States of America, the Argentine Republic, Brazil, and Chile" (article 9).[20]

These citations make it clear that the Rio Protocol allowed for the uncertainty of incompletely mapped terrain to permit adjustments in specific geographical situations (and anomalies) that might be found on the ground as demarcation proceeded as well as for navigation concessions on the Amazon and its northern tributaries, plus other possibilities within an additional bilateral treaty of commerce and navigation. As the parties to the conflict, Peru and Ecuador are denoted as the powers ultimately responsible for its resolution, but the obligation of the guarantors to assist is also explicit. The protocol, in sum, contains the necessary legal provisions for a solution to which both countries could ultimately agree.

Between 1942 and 1948, in fact, the Ecuador-Peru Boundary Commission, with guarantor help at several important junctures, was able to create a definitive demarcation of over 95 percent of the border without incident in accordance with the protocol's stipulations. Various "technical problems" which arose in the process were referred to the guarantors as specified in the treaty. Brazil, as guarantor coordinator, responded effectively and quickly in the six cases which came up, producing solutions that were accepted by the parties in five of them, including the area of the Cordillera del Cóndor between the Zamora and Santiago Rivers under the Braz Dias de Aguiar arbitral award of July 1945.[21]

An anomaly not foreseen in the treaty arose, however, after U.S. Army Air Corps mapping of the region was completed in 1946 and the surveys were turned over to authorities of both countries in February 1947.[22] This was the revelation by the aerial survey that the height of land which was to determine the border was not where the agreement had stipulated in one small section due to the presence of a previously uncharted river and mountain spur. As a result, this section of the border was not marked, and in 1948 the newly elected Ecuadorean government of Galo Plaza Lasso (1948–1952) ordered its demarcation team to cease activities. This area of the Cordillera del Cóndor and Cenepa River (totaling about 78 kilometers or 48 miles) of the Zamora-Santiago-Yaupi Rivers segment specified in the Rio Protocol was the primary focus of tension and the main locus of periodic outbreaks of hostilities until 1998.

Another interpretation of the dynamic between the parties, however, suggests a slowly changing position on the part of Ecuador with regard to a final and complete marking of the boundary.[23] This account draws from the George McBride reports in situ to note that Ecuador was quite aware of the existence of the Cenepa River as early as 1943, when a joint party of the Ecuador-Peru Boundary Commission followed it to its headwaters.[24] Subsequently, the account continues, Ecuadorean officials proceeded to secure a number of delays in the commission's work, which appear, with hindsight, possibly to have been designed to find some way to keep open Ecuador's chances for access to the Amazon in this sector.[25]

Ecuador rejected Peru's proposed solution and began building its legal case for the inapplicability of the protocol in the Cordillera del Cóndor region, as well as developing a diplomatic offensive for international support. In 1950 President Plaza Lasso withdrew Ecuador from the border demarcation commission. At the 1951 OAS meeting to consider the hemispheric response to extracontinental aggression, Ecuador's foreign minister raised the issue of intracontinental aggression to no avail.[26] The 1959 OAS conference of foreign ministers, scheduled for Quito, was postponed when Peru refused to attend because the host country included the dispute on the agenda. The return to international diplomacy was also accompanied by renewed border clashes.

In 1960 president-elect José María Velasco Ibarra (1934–1935, 1944–1947, 1952–1956, 1960–1961, 1968–1972) responded to Ecuador's inability to create an international coalition around the idea of inapplicability by raising the stakes further; he declared that, in addition to being inapplicable, the protocol was null because it was the result of Peruvian military aggression. Ecuador attempted to build a coalition around this idea in both the OAS (1959, 1965, 1980, and 1981) and the United Nations (1976, 1980, and 1991) but failed completely. By questioning a treaty negotiated after a war, the Ecuadoreans touched a sensitive nerve in the international community, which reacted by asserting the sanctity of international treaties. In response, Velasco Ibarra in his next term declared that "an honorable transaction" (sovereign access somewhere to the Amazon) could allow Ecuador to accept the protocol.[27]

Despite the unsettled issue, by the 1960s Ecuador had joined Peru as allies in regional foreign policy matters. The Andean Pact, created in 1969, seemed an ideal forum in which to replace concerns about territorial divisions with economic integration. Relations between the two countries improved markedly, and trade expanded.[28] Between 1970 and 1975 Ecuador and Peru signed a number of economic cooperation agreements, and economic cooperation accelerated after 1985. Alberto Fujimori (1990–2000) became the first Peruvian head of state to travel to Ecuador, offering various economic development proposals as well as the possibility of a free port for Ecuador on the Peruvian Amazon.[29]

Ecuador, however, entertained the idea that economic cooperation could be separated from its territorial claims.[30] It was not just the political elite that made this distinction between economic and territorial sovereignty issues. A public opinion poll conducted before the 1995 war demonstrated that Ecuadoreans could appreciate the need for greater economic relations with Peruvians but still distrusted them (49% of respondents believed that Peru was an "enemy" country and only 39% perceived it as friendly; see Appendix B).

Peru also refused to let economic cooperation affect its position on Amazonian sovereignty. Although the Andean Pact began to lose momentum in 1976 (partly because the less-developed Ecuador and Bolivia did not receive the benefits they hoped for), Peru rejected Ecuador's contention that progress on the border could provide the impetus for renewed progress among pact members.[31] Brazil's initiative for Amazonian cooperation and development seemed to offer Ecuador an opportunity to increase its de facto presence in the Amazon, but Peru short-circuited this effort in the final treaty signed with Brazil in 1978.[32]

In short, international diplomacy and economic cooperation did little to resolve Ecuador's Amazon problem. This failure led to a renewal of militarized clashes in 1977, just as democracy was in the process of returning to both countries—in Ecuador in 1979 and Peru in 1980. During the course of 1980 the stakes were raised, as Ecuador deployed forces into areas within that small section of disputed boundaries claimed by Peru. In January 1981 serious incidents occurred in what came to be called the Paquisha conflict, and Ecuadorean forces were quickly driven out of their outposts in the disputed area. Given its position on the Rio Protocol at the time, Ecuador called on the OAS to assist, while Peru recurred to the protocol guarantors.

The diplomatic solution that was worked out on this occasion, considering Ecuador's objections, was to use the OAS as the international mediating instrument with "friendly country" members (the four protocol guarantors) as its agents. These "friendly countries" oversaw a settlement in which the military forces of both parties moved to positions that they both recognized as being within the other's territory. Even though the protocol guarantors did succeed in restoring the peace under the OAS rubric, they were not successful in helping the parties to resolve the dispute itself, because Ecuador's diplomatic position did not permit the guarantors to have recourse to the provisions of the 1942 treaty.[33]

By the early 1990s internal political dynamics came to favor Ecuador, historically the weaker party in the border dispute. At that time Peru was undergoing the most profound domestic crisis in its modern history. Successive elected governments had failed to deal effectively with the multiple challenges of high foreign debt repayment obligations, internal economic problems, and a serious Maoist guerrilla threat from Shining Path.[34] The president of Peru elected in 1990, untested political neophyte Alberto Fujimori, inherited a country in crisis. He came to power amid a totally bankrupt bureaucracy, a cumulative inflation of more than 2 million percent over the five-year course of the first Alan García Pérez government (1985–1990, 2006–2011), which had impoverished much of the population, and an ongoing guerrilla war which was destroying Peru's infrastructure and killing thousands of its citizens. By the end of 1990 it had caused over $10 billion of direct damages—in a $40 billion economy—and ongoing political violence which had cost more than 23,000 lives.[35] While Ecuador was experiencing its own political and economic problems at the time, they were quite minor indeed by comparison with Peru's.

Given these domestic challenges, the Fujimori government from the outset pursued a regional diplomatic offensive designed to reduce the possibility of border problem flare-ups in order to be able to concentrate on internal concerns. The initiative included state visits and multiple conversations with presidents and high officials of Ecuador, Bolivia, and Chile in 1991 and 1992. The most important was Ecuador, given the continuing boundary disagreement. After another minor military infiltration within the disputed area at Pachacutec in 1991, this time by Peru, Peru's foreign minister, Eduardo Torres y Torres Lara, and his Ecuadorean counterpart, Diego Cordovez, attempted to fashion a "gentlemen's agreement" to establish a common security zone in the region.[36] Even though this effort foundered, others did not, including a state visit by President Fujimori to Ecuador, the first ever by a Peruvian chief executive. With the resulting reduction of tensions along this border, Peru could redeploy a significant portion of its military forces closer to centers of guerrilla activities in its pursuit of the counterinsurgency campaign against Shining Path, which was ultimately quite successful.[37]

The 1995 War

The precipitant of war in 1995 was the gradual redeployment of Ecuadorean military units into the disputed territory east of the Cordillera del Cóndor along the Cenepa River, beginning in 1991. These units constructed three heavily fortified bases on high ground in the area at Tiwintza, Cueva de los Tayos, and Base Sur, complete with significant artillery emplacements and protected by thousands of land mines. Although it seems to be the case that Peru became aware of this redeployment, no effort was made to respond to the Ecuadorean initiative until mid- to late 1994, when local Peruvian patrols on "friendly" visits to the Ecuadorean outposts issued warnings that they should withdraw.[38] Unlike the denouement of previous incidents, on this occasion little diplomatic activity took place before the outbreak of large-scale hostilities. This suggests a more significant role by the militaries of both countries in the 1995 developments and a corresponding marginalization of their diplomatic communities.[39]

When Ecuadorean forces did not respond to this "friendly" warning, as one account has it, local Peruvian forces took matters into their own hands, first by probing (in December 1994 and early January 1995) and then by rushing into a jungle confrontation for which they turned out to be ill prepared.[40] Outmanned and outgunned, they suffered casualties and were forced to withdraw (January 26, 1995). At this point the military forces of Peru mobilized and began to confront the already mobilized Ecuadorean military. However, they limited their fighting quite specifically to the disputed area in the Upper Cenepa valley. Over the course of the next five weeks, according to published reports, one hundred to three hundred casualties were inflicted, Peru lost nine aircraft, and in all the two governments expended an estimated $500 million.[41] As best we can tell, Ecuador remained firmly in control of at least the Tiwintza forward outpost in spite of published official announcements in Peru that "there was not an Ecuadorean soldier left on Peruvian soil."[42]

From a military standpoint, Peru suffered from significant disadvantages. Its supply lines were long and crossed over remote and difficult jungle terrain, which made moving troops and supplies to the frontier exceedingly difficult. Ecuador, in contrast, benefited from superior location, higher levels of military preparedness, and greater ease of resupply of its military outposts.[43] Unable to win militarily or to force a military stalemate, Peru went on the diplomatic offensive to remove Ecuador's forces through the Rio Protocol guarantor countries by declaring a "unilateral" cease-fire on February 13 and once again invoking their good offices.[44] Even so, Peru broke this cease-fire almost immediately: its military forces persisted in their attempts to get Ecuador out of the bases that its forces occupied. In fact, most of the casualties in the conflict occurred during the last two weeks of February. By the end of the month, however, skirmishes between the opposing forces had wound down for the most part. Although sporadic incidents continued for several months (with at least twenty recorded between May and September), the peacekeeping process under the Rio Protocol began to move forward.[45]

Postwar Peacekeeping under the Rio Protocol

Rapid recourse to the Rio Protocol and the assistance of the guarantors was quite unexpected, given Ecuador's long-standing opposition to its validity. It was made possible by the public request for an emergency meeting of the Rio Protocol guarantors by Ecuador's president Sixto Durán Ballén (1994–1998), to "inform them of the most recent border incidents . . . [and to ask] these countries for help in resolving [them]."[46]

President Durán Ballén and his civilian advisors were believed to have made the decision to invoke the good offices of the guarantors out of fear that Peru had the military capacity to push Ecuador's forces out of their forward positions once again, as in 1981, and perhaps take advantage of the situation to invade.[47] Ecuador's foreign minister Galo Leoro Franco confirmed his government's position that the protocol was indeed in effect but added that its "shortcomings" had caused incidents.[48] Ecuador's renewed willingness to work within the treaty's parameters was a significant shift back to a diplomatic position held during the 1950s— that the Ecuadorean government recognizes the existence (*vigencia*) of the protocol without accepting its validity (*validez*), due to the inability to execute (*inejecutibilidad*) one of its provisions.[49]

Peru immediately "acknowledge[d] and welcome[d]" Ecuador's declaration and also requested a convening of the guarantor countries to ask for their cooperation.[50] Guarantor country representatives accepted the request in an emergency meeting in Brasilia and soon issued statements urging cessation of provocations and separation of forces and noting a willingness to set up a mission to cooperate with Peru and Ecuador to find a solution to the situation.[51]

Beginning with the January 1995 Brasilia meeting, the peacekeeping process may be divided into three stages: (1) stabilizing the military situation on the frontier; (2) assisting the parties to articulate and specify the outstanding points of disagreement; and (3) assisting with negotiations between the parties to resolve the dispute. The guarantor representatives articulated five central principles: (1) unity of purpose, (2) military support for diplomacy, (3) a lead role for the parties, (4) legal recourse only, (5) setting their sights high.[52]

The basic document that provided the framework for pursuing guarantor objectives and observing these principles was the Itamaraty Peace Declaration, signed by the parties and the guarantors in Brasilia on February 17, 1995. Five of its six provisions were related to the military aspects of the conflict, including the establishment of a guarantor military observer mission to oversee implementation of the agreement. The sixth was diplomatic: "initiation of discussions between Peru and Ecuador to find a solution to the problems once compliance with [the other five] had reestablished sufficient mutual confidence."[53]

Ending the Fighting

Even though both parties and representatives of the guarantors had signed a peace agreement, serious armed confrontations continued in an area of the disputed section of the border. It proved necessary for the parties to reaffirm their commitment in Montevideo on February 28, when representatives of all the countries involved had gathered for the inauguration of the president of Uruguay. Only after this declaration did the hostilities cease, which then permitted the guarantors' proposed Ecuador-Peru Military Observer Mission (MOMEP) to be organized and to begin to carry out its duties.[54]

The next steps toward completion of the military stage of the peacekeeping process could now begin. This involved the separation of forces, withdrawal from the disputed area, and the establishment of a demilitarized zone. Over the next several months the MOMEP field mission was able to carry out these tasks, in spite of encountering several serious problems along the way.[55] The separation of Ecuadorean and Peruvian forces from the border areas began by the end of March; withdrawal of military units from the disputed area was completed, with MOMEP verification, by mid-May; and a meticulously drawn demilitarized zone of 528 square kilometers entered into effect in early August 1995.[56] By October diplomatic representatives of the parties and the guarantors and of Peru's and Ecuador's military representatives with MOMEP meeting in Brasilia could express "particular satisfaction" with the progress that had been made.[57]

Resolving the Conflict

Talks were initially hindered by Ecuadorean domestic politics. President Duran Ballén was in the last few months of his term and facing serious domestic opposition because of alleged corruption in his cabinet and a severe economic crisis. Interviews with Peruvian and Ecuadorean analysts suggest that the Fujimori government perceived that it had no one with whom to negotiate at the time. More importantly, all major social forces in Ecuadorean society advocated accepting nothing less than sovereign access to the Amazon. As a result, the dispute remained highly visible, with much political and diplomatic posturing.

In December 1995 Peru mobilized 6,000 troops on the border in response to Ecuador's purchase of four Kfir fighter bombers from Israel (with U.S. consent). Although military confidence-building measures occurred in the disputed sector during January and February 1996, the Peruvian negotiator arrived in Quito with copies of his book supporting Peru's interpretation of the 1947 negotiations.[58]

Negotiations stalled over a variety of issues, including Ecuadorean insistence on sovereign access to the Amazon, the need for its representatives to consult with Quito at every step in the process, national elections and internal political turmoil, and guarantor efforts to help build confidence and solid bases for resolution by the parties. Only after troop mobilizations in August 1998 produced another war crisis were the two countries able to make the concessions necessary for resolution. But they still could not tackle this most sensitive issue directly and asked the guarantors to propose a solution.[59]

The guarantors insisted that before taking up the task both congresses had to agree to abide by their decision. Within a week of getting such approval, the guarantors had their decision, along with a sweetener: a pledge of $3 billion in development aid from international financial institutions. Peru achieved a major aim: a technical commission named by the guarantors determined that the border lay along the height of land of the Cordillera del Cóndor. Yet Peru also had to make a difficult concession in exchange. Under the guarantor arrangement, Ecuador was granted perpetual use of a square kilometer of Peruvian territory at Tiwintza, the outpost that came to symbolize the 1995 war, to build a monument to its soldiers. And in a separate navigation treaty contemplated in the Rio Protocol, Ecuador also gained access to Peruvian waterways and to permanent port facilities on the Amazon River. As part of the agreement, Peru would build roads at its expense, connecting those facilities to the Ecuadorean rather than the Peruvian Pacific coast.

The Peruvian foreign minister, Eduardo Ferrero Costa, who had pursued a hard-line approach in the negotiations, resigned as soon as he learned that these were the terms to be offered by the guarantors.[60] After these terms were made public, riots broke out in Iquitos, the chief Peruvian city in the Amazon. Fujimori stood his ground, however. Ecuadoreans accepted the trade of a sovereign outlet to the Amazon that they had never possessed for guaranteed access, the promise of development aid, and the satisfaction that Ecuador had forced Peru to negotiate at least something in exchange for the military's battlefield success in the Cenepa.

Conclusion

As this summary description of the historical context, the 1995 war, and its eventual resolution makes clear, the differences between Ecuador and Peru over the boundary between the two countries were of long standing and deeply ingrained. They were resolved only after a brief but intense war in the disputed area and more than three years of on-and-off negotiations between the parties and the intense efforts by outside actors invoking the provisions of an international treaty agreement. In the next four chapters, we analyze the key factors that help to explain both why war broke out in 1995 and why resolution of the dispute could eventually be achieved. They revolve around considerations of power, domestic institutions, and leadership in each country, plus the critical role of hemispheric diplomacy that gradually brought the parties to a point where political and military leaders on each side could accept a solution.

Historical Time Line, Ecuador-Peru Border Dispute

1542	Francisco de Orellana Expedition to the Mouth of the Amazon from Cuzco and Muti via the Napo River.
1717	Separation of Viceroyalty of Nueva Granada (which included Audiencia of Quito) from Viceroyalty of Peru (voided in 1723 and reestablished in 1739).
1802	*Cédula* (edict) of the king of Spain separating most of the trans-Andean territory from the Viceroyalty of Nueva Granada and the province of Quito and transferring it to the Viceroyalty of Peru.
1822	Battle of Pichincha (May 22), securing independence of Ecuador (as part of Gran Colombia).
1823	Conclusion of boundary treaty with Peru by Joaquín Mosquera, commissioned by Simón Bolívar, establishing borders on the basis of *uti possedetis* as of 1809; but not ratified by congress of Gran Colombia.
1824	Battle of Ayacucho (December 9), securing independence of Peru.
1827	First request for United States mediation by Peru's foreign minister accepted, but belated arrival of U.S. representative (1829!) renders it moot.
1829	Battle of Tarqui: forces of Gran Colombia defeat Peru and rights to Guayaquil are reaffirmed.
1829	Treaty of Guayaquil, establishing borders as "the same as the former Viceroyalties of Nueva Granada and Peru before their independence"; not executed due to separation of Ecuador from Gran Colombia on May 13, 1830.
1830	Pedemonte-Mosquera Protocol, establishing the Marañón as the boundary between Ecuador and Peru, but never ratified by congresses.
1832	Pando-Noboa Treaty, recognizing present boundaries "until an agreement fixing the boundaries is concluded"; ratified by both parties.
1859	Peruvian occupation of Guayaquil in war with Ecuador.
1860	Treaty of Mapasingue, by which Ecuador recognizes territorial claims of Peru under the Cédula of 1802, but canceled by congresses of both countries in 1861.
1887	Bonifaz-Espinoza Treaty, by which Ecuador and Peru agree to submit their boundary dispute to the arbitration of the king of Spain, ultimately unsuccessfully.
1890	García-Herrera Treaty, which reaches a compromise on the borders by drawing a boundary approximating territories traditionally under the jurisdiction of each country; ratified by Ecuador but not by Peru; Ecuador revoked in 1894.
1904	Valverde-Cornejo Protocol, reviving the option of arbitration by the Spanish king, which produces recommendations provoking popular protests in both countries and a shift from arbitration to mediation and from Spain to the United States.
1910	U.S. mediation efforts, expanded to include Argentina and Brazil, which ultimately propose an Arbitration Tribunal at the Hague; Peru accepts, but Ecuador does not.

1924	Ponce-Castro Oyanguren Protocol, by which parties to meet in Washington to negotiate, submitting remaining differences to the U.S. president for arbitration; delayed, accepted by President Franklin Delano Roosevelt in 1934, and pursued beginning in 1936.
1936	Act of Lima, reaffirming commitment to 1924 agreement, maintaining boundary status quo "without recognition of territorial rights" in the meantime.
1936–1938	Inconclusive negotiations between parties in Washington, with good offices of the United States and efforts to expand to multilateral good offices by the Chaco War mediators.
1941	Peru-Ecuador War (July–September), resulting in a decisive defeat for Ecuador.
1942	Rio Protocol (January 29): treaty of "peace, friendship, and boundaries," signed by Peru and Ecuador, with the United States, Argentina, Brazil, and Chile signing as "guarantor" countries, approved by the congresses of Peru and Ecuador on February 26.
1942	Binational Ecuador-Peru Boundary Commission formed (June) and deployed to the field to place the border markers; technical differences submitted to guarantors for resolution in 1944 and 1945 and western boundary differences resolved.
1945	Arbitral decision on eastern boundary differences submitted by Braz Dias de Aguiar.
1947	Aerial mapping survey by U.S. Army Air Force, which turns over maps to the parties (February).
1948	Order from the Ecuador Foreign Ministry (September) to its members on the Demarcation Commission to stop work in the Cordillera del Cóndor, "since the map showed that there was no single watershed."
1960	Ecuadorean president José María Velasco Ibarra's declaration that "the Rio Treaty is null" (August).
1981	Outbreak of hostilities between Ecuador and Peru (January) in the disputed Cordillera del Cóndor area, mediated through the OAS by "friendly countries" (the guarantors).
1991	New border incidents which produce the "gentlemen's agreement" between the foreign ministers of Ecuador and Peru, quickly disavowed by Peru.
1992	First official visit ever to Quito by a Peruvian head of state by President Fujimori (January), followed by two more trips during the year.
1995	Major outbreak of hostilities in the disputed border area (January), producing a call by both parties for the good offices of the guarantor countries under the Rio Protocol.
1996–1998	Four formal meetings of Ecuador and Peru Foreign Ministers (January 1996–January 1998).
1998	Negotiations by commissions in guarantor capitals (February–July).

1998	Submission of guarantor-sponsored technical commission report that determines the border to be the Cordillera del Cóndor height of land as stipulated in the 1942 protocol (May).
1998	Active involvement of presidents Mahuad and Fujimori in seeking to resolve final obstacles (July–October).
1998	Final solution offered by guarantors; Peace Agreements signed by presidents Mahuad and Fujimori in Brasilia (October 26).

Source: Events through 1981 from Krieg, *Ecuadorean-Peruvian Rivalry in the Upper Amazon*; subsequent developments from newspaper accounts.

1995 Hostilities Time Line, Ecuador-Peru Border Dispute

January 9–11	Exchange of fire between Peruvian and Ecuadoean military patrols.
January 24	Ecuador recognizes Rio Protocol and asks guarantors for assistance.
January 26	Peru welcomes Ecuador's declaration and also asks aid of guarantors. Serious outbreaks of hostilities commence.
January 31	Guarantors meet in Brazil, invite Peru and Ecuador to participate, and both accept.
February 5	Peru accepts guarantor cease-fire proposal, Ecuador does not.
February 13	Peru declares unilateral cease-fire, which Ecuador accepts.
February 17	Parties sign Peace Accord of Itamaraty along with guarantors, but fighting continues.
February 28	Hostilities formally end with acceptance of both parties of the Montevideo Declaration, reaffirming validity of Itamaraty Accord.
March 10	Agreement on procedures signed in Brasilia by guarantors and parties.
March 30	Separation of Peruvian and Ecuadorean forces begins.
April 30	Forces of both sides largely withdrawn from disputed area (90 percent).
May 3–13	Withdrawal of all units from disputed area, with MOMEP verification, except for designated concentration points.
July 25	Establishment of a demilitarized zone by MOMEP, "without affecting the territorial rights of the parties to the conflict."
August 4	Entry into effect of a 528 square kilometer demilitarized zone.
October 5–6	Meeting in Brasilia of guarantor country officials with vice ministers of foreign relations of Peru and Ecuador, expressing particular satisfaction with progress.
November 17	MOMEP declaration noting satisfaction with progress in achieving a security accord for direct coordination between Peruvian and Ecuadorean military forces and with the absence of incidents.
December 27	Brief border incursion by Ecuador's forces, protested by Peru.

January 2, 1996	Peru expresses opposition to Ecuador's plan to acquire planes from Israel (Kfirs), with approval by a guarantor country, the United States, because of U.S.-made engines.
January 17–18	Lima meeting of Peru and Ecuador, with presence of guarantor representatives, to cover procedures for continuing search for a peaceful solution.
February 22–23	Quito meeting of Ecuador and Peru, with guarantors, to advance the process, including agreement to submit list of remaining substantive differences in achieving an accord.
March 6	Public release of the lists of remaining differences. Peru wants final drawing of the boundary line; Ecuador continues to note the inapplicability of the protocol to one area and requests sovereign access to the Marañón-Amazon. Parties agree to continue discussions in the near future, under the auspices of the guarantors.
June 18–19	Buenos Aires meeting of Ecuador and Peru to continue procedural discussions, with the presence of guarantor representatives.
October 28–29	Santiago meeting of Ecuador and Peru to complete procedural discussions, with the presence of guarantor representatives. Plan to meet in Brazil to begin substantive talks on December 20, 1996, postponed—first by the hostage crisis in Lima in mid-December and then by the replacement of Ecuador's president by the congress on February 6, 1997.
April 15, 1997	Official negotiating commissions designated by Peru and Ecuador meet with guarantor representatives in Brasilia to implement the Santiago Agreement of October 29 by beginning substantive discussions on remaining impasses.
1998	Ecuador's vice minister of foreign affairs omits reference to "sovereign" access to the Amazon (January).
1998	Commissions negotiate in guarantor capitals (February–July).
1998	Guarantor-sponsored technical commission report submitted that determines the boundary to be the Cordillera del Cóndor height of land as stipulated in the 1942 protocol (May).
1998	Presidents Mahuad and Fujimori become actively involved in seeking to resolve final obstacles (July–October).
October 26, 1998	Guarantors offer final solution and presidents Mahuad and Fujimori sign peace agreements in Brasilia.

Sources: Gabriel Marcella, *War and Peace in the Amazon: Strategic Implications for the United States and Latin America of the 1995 Ecuador-Peru War*; Col. Glenn R. Weidner, "Peacekeeping in the Upper Cenepa Valley: A Regional Response to Crisis"; newspaper accounts; and Francisco Carrión Mena, *La paz por dentro, Ecuador-Perú: Testimonio de una negociación.*

3 PRESIDENTIAL DECISION MAKING
The Institutional and Personal Context

Institutions constrain political leaders' behavior and policy choices. By establishing precedents and constitutional guarantees, institutions lend legitimacy to certain themes while denying it to others. They also affect, directly and indirectly, who can participate in the domestic political game as well as how.[1] The institutional context is not determining, but it does affect which interests matter and what leaders need to do in order to maintain the support of their policy constituencies. The leaders' skill in devising new responses to their constituencies' interests (innovativeness) and their willingness to try them out (risk acceptance) help us understand the degree to which leaders are likely to maneuver and make policy on the margins of those institutional constraints. The task of this chapter is to analyze the relevant constraints upon Ecuadorean and Peruvian leaders and categorize the leadership characteristics of the five presidents (Sixto Durán Ballén, Abdalá Bucaram, Fabián Alarcón, and Jamil Mahuad in Ecuador, Alberto Fujimori in Peru), to establish the context within which the policies that produced war in 1995 and peace in 1998 can be understood.[2]

The distinction between democratic and nondemocratic institutions is not sufficient to capture the variation in these important constraints upon decision making. Ecuador democratized in 1979, Peru in 1980; both countries were democracies at the time of the major conflict in 1981 and when full-scale war broke out in 1995. If anything, one might postulate that democracy was weaker in both countries in 1998 when peace was negotiated than earlier when war developed.[3] In order to understand this counterintuitive result,[4] we have to delve more deeply into the way in which these democratic institutions actually functioned.

The first section of this chapter analyzes electoral accountability and the party system in Ecuador and Peru. Political parties constitute the chief means by which citizens aggregate their votes in representative democracies; thus it is fundamental to understand the ability of Ecuadorean and Peruvian parties to link citizens and political institutions.

The second section examines the executive's relationship with the legislature and Supreme Court, which are expected to constitute institutional constraints on the executive. These relationships provide us with an appreciation of the actual space available to the Ecuadorean and Peruvian presidents to make policy decisions on their own terms.

The third section discusses civil-military relations in the two countries, in recognition of the active and often legitimate though still mainly informal role that the military has historically played in both countries. Because military coups are virtually never undertaken without significant civilian support,[5] we are interested in discovering the extent and limitations of military influence in policy making in each of the two countries.

The fourth section deduces the innovative and risk acceptance characteristics of the five leaders involved in the war and subsequent peace. The conclusion summarizes our argument about the institutional constraints and leadership characteristics that operated when war broke out in 1995 and peace was signed in 1998.

Electoral Sanctions and the Party System, 1995–1998

The link between voters and their political representatives is a fundamental determinant of government policy. The electoral rules and party system are the key characteristics of the voter-politician link that we must understand. Among electoral rules are the ballot procedures and term limits; each affects the accountability of elected politicians to their constituencies. There are important trade-offs between accountability and representativeness, and the number of parties affects this relationship. More parties mean more diversity of opinion in the legislature and thus increased representativeness. However, more parties also make it more likely that legislators will become less accountable to the electorate because they can blame other parties for policy failures, thus distancing themselves from the most obvious criterion that their constituencies can use to evaluate their performance.[6] Party strength also impacts both legislative behavior and accountability to the electorate.

Ecuador

Ecuador's democratic polity in the 1990s was a presidential system with a unicameral legislative assembly that functioned under three different constitutions passed in 1979, 1996, and 1998. The president was elected every four years in national elections under a runoff system: a second round was held with the top two vote getters if no one received at least 50 percent in the first round. The eighty-two assembly members were elected through two processes: twelve via a national list (four-year terms), and seventy through provincial lists (two-year terms). Seats were allocated on a proportional representation basis. Voters chose among "closed lists," selecting parties rather than the candidates themselves; *ceteris paribus*, this procedure should have strengthened parties but diminished the link between legislators and the electorate. Reelection was not permitted for the legislature or the presidency, further limiting accountability of the government to the electorate.

In 1994 the electorate began to gain more influence over politicians. The constitution was amended in that year to allow presidential reelection once, but not consecutively, as well as unlimited congressional reelection; the first reelections were held in 1996. Legislative elections were concurrent with the first round of a presidential election, with midterm elections for members elected on provincial lists, thereby providing some ability for voters to signal their views about the president's performance by voting for or against parties supporting the chief executive.

When constituencies have very little influence over their elected representatives, politicians are free to vote with their parties (a "partisan" vote) or in line with their individual policy preferences (a "personal" vote). A legislature dominated by personal voting makes it difficult for controversial legislation to pass because of "buck-passing." Legislators do not want to assume responsibility for taking a controversial position and are not punished for failing to adopt legislation to deal with the controversial issues. Strong parties could overcome this policy-making problem by imposing discipline on their members.

Table 3.1. Party Switching Rates in Ecuador (1979–1998)

Legislature	Party Switching Rate (%)	Independent Legislators (%)
1ˢᵗ Legislature (elected in April 1979)		
April 1979–August 1979 transition	0	0
August 1979–August 1980	28	3
August 1980–August 1981	16	14
August 1981–August 1982	10	7
August 1982–August 1983	14	3
2ⁿᵈ Legislature (elected in January 1984)		
January 1984–August 1984 transition	8	8
August 1984–August 1985	8	11
3ʳᵈ Legislature (elected in June 1986)		
June 1986–August 1986 transition	1	1
August 1986–August 1987	10	10
4ᵗʰ Legislature (elected in January 1988)		
January 1988–August 1988 transition	1	1
August 1988–August 1989	1	1
5ᵗʰ Legislature (elected in June 1990)		
June 1990–August 1990 transition	0	0
August 1990–August 1991	10	10
6ᵗʰ Legislature (elected in May 1992)		
May 1992–August 1992 transition	3	1
August 1992–August 1993	18	18
August 1993–September 1993	9	27
7ᵗʰ Legislature (elected in May 1994)		
May 1994–August 1994 transition	0	3
August 1994–August 1995	21	17

8ᵗʰ Legislature (1996)		
1996	18	13
9ᵗʰ Legislature (1998)		
1998	18	9

Sources: 1979–1994: Proyecto CORDES Gobernabilidad (Quito: CORDES, 1997); 1996 and 1998: Andrés Mejía Acosta, "La reelección legislativa en Ecuador: Conexión electoral, carreras legislativas y partidos políticos (1973–2003)," Ecuador Debate 62 (2004), 251–269; 1996 and 1998: Análisis Semanal (Quito).

Party behavior in Ecuador, however, has been plagued by a lack of discipline and factionalism. The inchoate Ecuadorean party system resulted from several factors: the multiplicity of parties due to low voting threshold requirements for legal status, the presence of midterm elections from provincial lists, the weakness of party label identification by the electorate, an antiparty mentality fostered by politicians themselves, and term limits until 1996.[7]

A good indicator of the lack of party discipline is party switching on the part of legislators, a phenomenon known in Ecuador as *cambio de camisetas*. Corporación de Estudios para el Desarrollo (CORDES) calculated the percentage of legislators who bolted their parties during congressional transitions (that is, the period between the date a legislative election is held and the date a newly elected legislature is inaugurated) and one-year legislative periods.[8] Table 3.1 displays party switching rates and the percentage of independent legislators in Ecuador in the 1979–1998 period. The existence of independent legislators complicates legislative bargaining because their votes require negotiations at the individual rather than group level. When combined with party switching rates these two indicators provide strong support for the view of Ecuador's party system as dominated by "floating" politicians.[9]

Party switching does not just weaken parties; it can also cause them to proliferate. Another criterion that can help us understand the congress's ability to function effectively and be accountable to the electorate is the effective number of parties. Not all parties are relevant, and, as we have seen, many seats are often held by independents. Consequently, we cannot just count the parties represented in the legislature. Appendix A elaborates on the methodology that we use to arrive at the effective number of parties. Since high rates of party switching in Ecuador affect legislative fragmentation by inflating some parties and deflating others in a single legislature, we calculate the effective number of legislative parties per year rather than per legislature.

The average effective number of parties for 1979–1998 was 6.42. Ecuador ranked among the most fragmented legislatures in the world, together with Belgium, Brazil, Poland, and Russia. The standard deviation was also high (1.45), but the most striking feature of the Ecuadorean experience is that the highest numbers were obtained in the middle of legislatures, not after elections. This pattern indicates that fragmentation was mostly determined by elite decisions rather than by voters' preferences, which confirms the inability of voters to hold their legislators accountable. Moreover, variation in fragmentation in a single legislature—owing to high rates of party switching—indicates that voters did not determine the composition of the congress, let alone the composition of governing majorities. Three midterm elections were held (1986, 1990, and 1994); the one scheduled for 1998 became a presidential election to replace interim president Fabián Alarcón, who had taken office after president Abdalá Bucaram and vice president Rosalía Arteaga were forced from office in 1997. But the party switching phenomenon persists, thereby diminishing the ability of midterm elections to serve as a means of holding legislators accountable.

Peru

Since redemocratization Peru has had two constitutions, one in 1979 during the transition from military to civilian rule and another in 1993 after President Fujimori closed congress in 1992 and a constituent assembly was elected. Under both constitutions, elections for the legislature follow the d'Hondt formula of proportional representation (PR) with an open list. Of the multiple PR systems, d'Hondt is the "least proportional and systematically favor[s] the larger parties."[10] Reelection to the legislature was permitted. The open list format and the possibility of reelection increase the legislators' accountability to their constituencies. Under the 1979 Constitution, the legislature was bicameral, but the 1993 Constitution responded to popular sentiment and made it unicameral.[11]

The constitutions differed with respect to reelection of the president. Under the earlier constitution a president could be reelected multiple times, but only after sitting out a term, while the 1993 Constitution permitted an immediate reelection, but the incumbent then had to sit out a term before running again.[12] The 1979 Constitution made outgoing presidents senators for life (senadores vitalicios). In general, presidents who will be seeking reelection should be highly accountable to the electorate;

Table 3.2. Effective Number of Parties: Peru

Year	Chamber	Effective Number of Parties
1980	Senate	3.43
1980	Deputies	2.50
1985	Senate	2.60
1985	Deputies	2.18
1990	Senate	4.04
1990	Deputies	4.14
1995	Unicameral	2.92
2000	Unicameral	3.78

Source: Calculations by Lydia Tiede based on data from IFES election guide 2000 (www.electionguide.org/results .php?ID=650/, accessed July 19, 2007), using the formula in Appendix A.

the one-term layoff required under the old constitution would have diminished but not eliminated that accountability if the incumbent had hopes of running again in the future. Because the new constitution abolished the "senator for life" provision and permitted immediate reelection, electoral constraints on the Peruvian president increased after 1993. Like the Constitution of 1979, however, its 1993 counterpart did not provide for midterm elections, thereby continuing to offer no opportunity for the electorate to judge the performance of the incumbent administration until the end of its term.

Peru's party system instability rivals that of Ecuador. Among the twelve Latin American countries studied by Scott Mainwaring and Timothy Scully, Peru experienced by far the greatest levels of congressional electoral volatility ("net change in the seat shares of all parties from one election to the next"), while Ecuador was fourth highest.[13] Peru's effective number of parties was significantly lower than Ecuador's average of 6.42, however, though it spiked upward in the crisis context of 1990 (see Table 3.2).

In summary, the electorate confronted serious obstacles in both Ecuador and Peru in holding their national elected officials accountable. But both the 1993 Constitution in Peru and the 1994 constitutional reforms in Ecuador increased electoral accountability on the eve of the 1995 war. By the time the peace treaty was ratified in 1998, electoral accountability was in the process of weakening in both Peru and Ecuador. In Peru some citizens showed their displeasure by protests and riots, particularly in the jungle city of Iquitos. In Ecuador congress had removed an elected president in 1997, giving way to an interim government until new presidential elections in mid-1998.

Executive Relations with Legislative and Judicial Branches

Party system attributes, while necessary to our analysis, are insufficient to provide a useful understanding of the dynamics of a democratic polity, particularly of presidential systems like Ecuador and Peru. As Scott Mainwaring and Matthew Shugart stress,[14] the key to a proper understanding of Latin American presidential regimes is to be found in the interaction between the institutional design of the system of government and the party system.

Ecuador

In Ecuador's political system,[15] presidents had both proactive and reactive powers in their relationship with the legislature. Chief executives could propose legislation to congress for consideration; if they declared it "urgent" (economic legislation only), congress had up to fifteen days to act on it before it automatically became law. This provision is akin to presidential decree power if the legislature is fragmented and cannot agree on what to do about the issue on which the president has proposed legislation.[16] Moreover, the president could call national referenda on questions, "which, in his judgment, are of great importance to the state," in particular on constitutional amendments and on the ratification of international agreements that have been rejected by congress.[17]

The high levels of fragmentation in a single legislature (owing to party switching and a large number of effective parties) suggest that party elites rather than voters were more important in determining the composition of the congress as well as the composition of governing majorities. One result was that the president's party never dominated congress: legislation required a coalition among parties to be enacted.

Additionally, the chief executive could appoint and remove cabinet ministers without restriction. Members of the presidential cabinet, however, could be censured by a majority of congress for loosely defined "violations committed in the fulfillment of their office." From 1979 to 1998 almost 10 percent of ministers were censored.[18] A minister censured by congress could not serve in any governmental role during the rest of that presidential term. The president could declare a state of emergency, which congress might revoke but need not act to uphold, in case of foreign invasion or internal disturbance. The chief executive could also decree the collection of taxes in advance, impose censorship, and suspend constitutional guarantees in cases of declared national emergency. In addition, the president managed a "reserve fund" which was not subject to congressional authorization or oversight, except in the case of allegations of corruption.

The president did not have a veto on the budget, but on all nonbudgetary legislation the legislature could not reconsider any bill vetoed by the president for one year. Congress could, however, call for a binding national referendum on the vetoed bills. If the president's objection related only to a specific part of a bill, congress could either rectify the section (resulting in promulgation of the bill) or override the veto by a two-thirds vote.

Congress also demonstrated that it was willing to make cynical use of constitutional provisions to drive out the president; in February 1997 congress voted to remove controversial President Abdalá Bucaram on grounds of "mental incompetence" despite never having subjected him to any medical tests. The Supreme Court subsequently ordered his arrest for corruption, the military considered trying him for treason, and congress banned him from running for public office, an action ratified by a national referendum.[19]

Members of the Supreme Court were appointed by congress for six-year terms and were eligible for reelection. Judgeships on the Supreme Court and Tribunal Courts were parceled out to the political parties in proportion to their representation in congress.[20] In actual practice, however, when congress was elected and the distribution of party strength changed every two years, the Supreme Court was "renewed" by replacing a few judges. A major change occurred every four years when a new president was elected and supporters in congress sought to provide a more compliant court. A 1997 plebiscite indicated that the public did not want congress to "politicize" the Supreme Court, so congress removed all thirty-one judges and appointed new ones.[21] The attorney general was also appointed by congress, and all judicial appointees were subject to possible congressional censure. Under the 1998 Constitution, Supreme Court justices finally achieved lifetime tenure. Appointment of new judges was limited to those with a judicial career and required a two-thirds vote of the existing court.[22] These two provisions produced a significantly more independent Supreme Court in Ecuador than had been the case before 1998.

Constitutional amendments could be proposed by legislators, by the president, by the Supreme Court, or by referendum. Approval of amendments required the consent of two-thirds of deputies in two separate rounds. A president who totally or partially rejected an amendment approved by congress could submit it to a referendum. The president could also call a referendum if a proposed amendment had been partially or totally rejected by congress.

Peru

The Constitutions of 1979 and 1993 strengthened presidential power, because stalemates between the legislative and executive branches of government were perceived by the respective constituent assemblies to have been a major cause of the coups of 1968 and 1992.[23] Nevertheless, formal legislative constraints on the executive continued to be important, so the 1993 Constitution should be seen as reinforcing a trend rather than as a break with the past.

Congress could censure a minister for noncriminal reasons. A minister who was censured was obliged to resign. Under the 1979 Constitution, if congress censured three ministers the president could close the lower chamber and call for new elections within thirty days; under the electoral rules operative at the time, this threat of dissolution provided a mechanism to keep congress from the indiscriminate use of censure. Dissolution could not be invoked in the last year of a president's term or during a state of siege (during a foreign or civil war). If elections were not held within the allotted time, the dissolved chamber reconstituted itself and dismissed the president's cabinet; no member was able to resume a cabinet position during that presidential term. The senate, however, could not be dissolved under any circumstance.[24]

Similar constraints on the executive during the process of congressional dissolution prevailed in the Constitution of 1993. Congress elected a permanent commission, made up of members of the parties represented in the legislature in proportion to their numbers. The permanent commission had the power of impeachment, with congress functioning as the jury. The senate was eliminated in the new constitution in favor of a single legislative body; now it was the permanent commission that could not be dissolved. The permanent commission and the new congress were empowered to examine any decrees issued by the executive during the intervening period.[25] Under both constitutions, a simple majority could overturn a presidential veto of legislation.[26]

The decree power of the executive was strengthened somewhat in 1993, although earlier presidents Fernando Belaúnde Terry and Alan García Pérez had both utilized their decree powers extensively (issuing 2,086 and 2,290 decrees, respectively).[27] Both constitutions stipulated that congress could delegate decree power to the president for specified matters and periods and that congress could overturn a decree.[28] After 1993 the executive could decree a state of emergency or siege, but not on matters that "the permanent commission cannot delegate." States of emergency (during periods of internal disorder) only required notification of congress and could not exceed sixty days without a new decree. States of siege (during wars) could not exceed forty-five days; congress had the right to convene, and any extension of the state of siege required congressional approval.[29]

The further expansion of executive powers in the Constitution of 1993 strengthened President Fujimori's hand in pursuing his policy objectives, but he was also aided substantially by gaining a legislative majority of his supporters in both the 1993 and 1995 congressional elections. The combination made it virtually impossible for opposition parties to impede President Fujimori's progressive usurpation of democratic procedures in the late 1990s, culminating in the fraudulent election of 2000.

Given decree powers, the structure of political parties, and the proscription against immediate reelection, presidents Belaúnde and García may have acted under fewer institutionalized constraints than President Fujimori under the new constitution. Even though President Belaúnde's coalition of center-right parties controlled congress, he pursued his agenda via decrees. He also knew that at his age (seventy-eight at the time of the relevant election) reelection after an intervening term was out of the question. President García's party also had a majority in congress, yet he also still made extensive use of his decree powers. Neither president made important changes in his governing program even in the face of precipitous declines in approval ratings. President Belaúnde ended his term with a rating in the low twenties and García with a rating in the teens. President Fujimori began to experience a fall in his own popular approval ratings in 1998; given the plan for mounting a reelection campaign in 2000, this may well have been a major factor in his government's efforts to manipulate the democratic process in his favor. An additional source of concern was the high approval ratings at the time of the mayor of Lima, his most likely rival at that point for the 2000 elections.

Institutional constraints had proven utterly unable to prevent President Fujimori from closing congress and purging the judiciary in 1992. His ability to override these constitutional constraints was fundamentally a result of the decline in the legitimacy of both congress and the judiciary in the eyes of the public, the willingness of the military to support the *autogolpe*, and the public's preoccupation with hyperinflation and Shining Path's guerrilla war. Immediately after the coup, 71 percent of those polled approved of closing congress, 89 percent approved of restructuring the judiciary, and 86 percent believed that Fujimori should remain president.[30]

Nevertheless, President Fujimori was constrained by the immediate response to his suspension of democracy when the Organization of American States (OAS) met in emergency session in Nassau, Bahamas, under the provisions of the 1991 Santiago Resolution (OAS Resolution 1080) to request the restoration of democratic procedures. He agreed to do so within a year and soon called for the election of a constituent assembly (Congreso Constituyente Democrático) to write a new constitution and then serve as Peru's legislative body until the next scheduled elections in 1995. Given his public support at the time (enhanced by the dramatic capture of Shining Path's leader, Abimael Guzmán Reynoso, in September 1992), it came as no surprise that his supporters captured an absolute majority in the November 1992 election (forty-four of eighty seats). The new constitution, which increased the chief executive's public policy prerogatives among other significant changes, was narrowly approved in a referendum a year later. The restoration of democratic procedures under the Constitution of 1993 had the additional effect of increasing President Fujimori's domestic institutional constraints as well as those derived from his need to respond to his constituencies. His coalition also won a narrow majority in the 1995 congressional elections in the context of success against Shining Path, the taming of hyperinflation, and the restoration of economic growth, so he continued to have the support he needed in congress to pursue his ongoing policy agenda. And that time the opposition to Fujimori had popular support. In short, in 1995 Fujimori would have found it difficult to justify to the country his overthrowing of the constitutional system when terrorism and hyperinflation had both been defeated, his party controlled Congress, and he was behind his rival in some polls.

Civil-Military Relations

During periods of great social, political, or economic change, institutional rules help determine whether civilians support the military's demand for the restriction of the political rights of those who question the path on which the nation finds itself. Restrictions on civilian dissent imply an increase in the police power of the state, which directly affects the position of the military in society.[31]

Ecuador and Peru carried out their transitions to democracy during the 1980s with a civil-military relationship best characterized as *parallel spheres of action*.[32] Politicians were charged with developing national wealth and internal stability. The military, in turn, legitimately monopolized every issue related to security, because such a division of labor was incorporated into the terms of the transition.

Ecuador

The Ecuadorean military crafted the transition to democracy in the 1970s in numerous ways to moderate civilian conflict (including proscribing the candidacy of populist Assad Bucaram, uncle of the subsequently impeached President Abdalá Bucaram), strengthen the civilian government's ability to govern, and limit civilian "meddling" in the military organization.[33] Civil-military relations in Ecuador provided the military with considerable formal and informal autonomy. The Constitution of 1979 gave the military a role in the social and economic development of the country, linking these to national security. Congress had the responsibility to exercise oversight on defense policy; but civilians lacked expertise on defense matters, thereby making it difficult for effective oversight to occur. The military was guaranteed a share of revenue from petroleum exports, thus further reducing civilian control of the military budget. It was a treasonable act to defame the armed forces.

The armed forces were the most respected institution in the country during this period, far outdistancing congress and the presidency.[34] The weight of such popular support further increased the influence of the military in politics. Some elements on the right of the political spectrum repeatedly called for military intervention against center-left governments. When it appeared that Abdalá Bucaram might win the 1988 presidential elections, opinion polls demonstrated majority support for a military coup to prevent him from assuming office. Some civilian sentiment for such a move still existed when Bucaram won the 1996 elections. President León Febres Cordero (1984–1988), in particular, sought to use the military in his disputes with congress and numerous social forces.

In spite of high levels of popular support, however, the military did not dominate the government. Ecuador is led by civilians, and the military prefers it that way. In the nineteen years of democracy preceding the

1998 peace agreement, only one military coup was attempted (by the air force high command against Febres Cordero in 1986), and it was quashed by the army itself. The army even stood by when congress acted on the margins of the constitution and impeached Bucaram in 1997, bypassed the vice president, and installed its own leader in the presidency.[35] Demonstrators' demand for the military to oust Mahuad in 2000 and the general public's acceptance of his forced replacement even when the coup against him failed illustrate that the 1997 episode of civilian recurrence to unconstitutional means of unseating leaders cannot be dismissed by citing the peculiarities of President Bucaram. As this event demonstrates, the military had sufficient popular backing to weigh in on a dispute and affect the outcome on the margins.

Peru

The Peruvian president has the initiative on national security policy, but operational control is largely in the hands of the military. In 1981 Belaúnde first tried to use special police forces (Sinchis) to deal with the Shining Path guerrillas in Ayacucho, but they were withdrawn after several months due to a number of incidents of bad behavior that outraged the local population. As the guerrilla threat increased in 1982, Belaúnde (with congressional approval) declared seven provinces in the Ayacucho area to be an Emergency Zone, which suspended some constitutional guarantees there and put the affected region under the political control of the local military base's commanding general. President García initially attempted to strengthen civilian control over the armed forces by creating a Ministry of Defense and developing his own paramilitary and intelligence agents operating out of the Interior Ministry. But his decision to rely on the army and the Republican Guard to put down a prison revolt by Shining Path inmates in 1986 and his attempt to promote Alianza Popular Revolucionaria Americana (APRA, American Revolutionary Popular Alliance) sympathizers within the officer corps short-circuited that effort.[36]

Under President Fujimori, the Peruvian civil-military relationship evolved into a *dominant-subordinate relationship* in which Fujimori, a civilian, was dominant. In this type of relationship the dominant partner identifies threats and designs appropriate responses. These actions are based on the dominant partner's interests and perceptions. Fujimori's coup of 1992 resolved the problem of civilian disagreement about the appropriate role of the military in periods of political strife between the legislative and executive branches in favor of the president.

President Fujimori made a bargain with one of the competing groups of officers to garner the institution's support for his government.[37] Some analysts see the military as severely constraining Fujimori, on the basis of his 1995 grant of blanket amnesty to the military for any human rights violations carried out in their fight against internal subversion or narcotrafficking as well as his 1992 announcement that military courts would have jurisdiction over trials of civilians accused of crimes against national security.[38] However, Argentine President Carlos Menem, whose antimilitarist credentials are widely acknowledged,[39] also extended an amnesty to the military in an effort to end what he saw as a corrosive human rights debate.[40] In President Fujimori's case, he shifted responsibility for trials of alleged terrorists to the military courts because the civilian judicial system had completely broken down in the face of long delays and systematic threats and assassinations of judges by Shining Path.

Furthermore, he kept his original choice of Gen. Nicolás Hermoza de Bari as chief of the Joint Command of the Armed Forces even after his retirement. This was not only unprecedented but was opposed by the military from the beginning because Hermoza did not meet the established merit requirements for the position. In addition, during the 1995 war President Fujimori decided to go to Tiwintza, accompanied by the press, against military advice.[41] He also removed Hermoza after the general claimed credit for the successful operation in April 1997 against the Movimiento Revolucionario Tupac Amaru (MRTA, Tupac Amaru Revolutionary Movement) guerrillas who were holding the Japanese ambassador and others hostage and opposed the diplomatic settlement with Ecuador as it was emerging in August 1998. In short, President Fujimori retained the initiative with the military, made a clear distinction between operational and policy questions, and was willing to intervene in operational questions when it affected his political program.[42]

Leader Characteristics

The personal qualities that chief executives bring to bear in the decision-making process can be and often are crucial in determining policy outcomes. Table 1.5 in Chapter 1 illustrates the relationship between innovation and risk acceptance and the institutional constraints under which leaders operate. As that table notes, the innovator pushes new ideas, while the risk taker tries to alter the institutional status quo. An innovator who is not a risk taker pushes new approaches only as far as the institutional context allows; for a risk taker who is not an innovator, institutional adjustments are vehicles for old ideas. A chief executive who is neither sticks to traditional policies within existing institutions and procedures.

In assessing the characteristics of Ecuador's and Peru's leaders, we cannot use the policy decisions relating to the 1995 war and the tortuous road to the 1998 peace to determine the degree to which those policies can be explained by their personal characteristics. The presidents of both Ecuador and Peru confronted major domestic political and economic challenges not related to the conflict and its resolution, however, so we can use their responses to ascertain whether they were innovators and risk takers.

President Sixto Durán Ballén (1992–1996: president during the 1995 war) was a conservative and initially adopted standard conservative policies to deal with the economic crisis: budget cuts, an increase in prices of government-provided goods (energy prices and petroleum derivatives), decreases in state employment levels, and exchange rate devaluations. This was the decade of the Washington Consensus promoting market economies, so privatization, minimizing government regulation, and promoting free trade pacts with neighbors (but not with Peru) were also standard responses in the Durán Ballén administration,[43] although a number of its fiscal reforms were ahead of their time. In the Ecuadorean political context, creating a new political party (Partido Unido Republicano [PUR, United Republican Party]) to seek office is not particularly innovative (Bucaram and an indigenous group did so as well), and Durán Ballén's privatization efforts were very cautious.

Furthermore, even though Durán Ballén's party had only 10 percent of congressional seats at the beginning of his term (which dropped to less than 5 percent after the midterm elections), congress still passed virtually all of his reforms, even those he initiated under his executive powers. This provides another indication that Durán Ballén was at best only a moderate innovator during his term.[44]

Abdalá Bucaram (August 1996–February 1997) was mercurial in temperament; his administration, short as it was, was also extremely corrupt. During his electoral campaign, his competitors labeled him "El Loco" (the crazy one) because of his tendency to drink heavily, dance wildly, and sing boisterously; Bucaram gleefully accepted the nickname. Once in office, he advanced fiscal and monetary proposals that were rejected by congress because they "lack[ed] political and economic credibility."[45] Even though the parties in congress were all turning against him and calls for the military to oust him were increasing, Bucaram did not moderate his behavior. We can thus identify him as a recklessly innovative and risk-taking leader.

Fabián Alarcón (interim president February 1997–August 1998) was very cautious and moderate during his brief term. While it was clear that he was in office only until new presidential elections could be held, Ecuador continued to experience serious economic and social challenges, including the rise of a new political party to represent the long ignored indigenous population. An innovator and risk taker would have attempted to push the limits of his interim status to promote what he thought were much-needed changes; Alarcón did not.

Jamil Mahuad (August 1998–January 2000) began his administration with moderate fiscal and financial reforms that garnered the support of congress, where his party controlled only 35 percent of the seats. But as the economic crisis heightened and then spun out of control, particularly in the banking sector, Mahuad pushed harder on economic reforms. After a run on the banks, he initially responded by making the banks responsible for the credits they had issued and then froze the accounts of depositors. Mahuad had sought support in congress from the right, but when that failed to materialize he successfully bargained with the left. However, massive and constant national protests continued through the second half of 1999.

Convinced that radical measures were necessary, Mahuad began to dollarize the economy, provoking his overthrow in January 2000 by a coalition of indigenous groups and some military officers. Although the coup attempt was put down, Mahuad was forced to resign in favor of his vice president. On balance, Mahuad's short-lived presidency provides sufficient material to characterize him as innovative and risk acceptant.

Turning to Peru, Alberto Fujimori (1990—November 2000) made a number of innovative and risky public policy decisions throughout his decade in the presidency. Upon taking office, he promoted a dramatic economic liberalization that caught his electoral supporters by surprise, angered many of them, and had little support in congress–but he went forward anyway. In 1992 President Fujimori responded to what he said was congressional opposition to his domestic policies by closing congress and suspending the 1979 Constitution. After being pressed by the international community to restore democracy, he was also intimately involved in writing a new constitution designed to legitimate market-oriented reform and neo-liberalism while also creating a unicameral legislature and enhancing some presidential powers.[46]

Table 3.3. Characteristics of Peruvian and Ecuadorean Leaders

Risk Taker/Innovator	Risk Taker/ Not Innovator	Innovator/ Not Risk Taker	Not Risk Taker/ Not Innovator
Mahuad and Fujimori strongly so Bucaram recklessly so	None	Durán Ballén moderately innovative	Alarcón

On the border dispute, Fujimori seized upon a Militarized Interstate Dispute (MID) in 1991 to push for a definitive settlement, although he was criticized in congress for not dealing harshly with Ecuador.[47]

Between 1991 and 1993 he offered Ecuador a package linking economic development projects, a free port on the Amazon, reciprocal security measures, and arms limitations along the border in exchange for a border demarcation specified by the 1942 protocol. His trip to Ecuador to offer details on the economic proposals represented the first such visit by a Peruvian president; he would go three times in all.

Table 3.3 provides a summary of the leadership characteristics of these heads of state of Ecuador and Peru in terms of innovating and risk taking.

Conclusion

The institutional context of democracy has important similarities and differences in Ecuador and Peru. Politicians are voted out of office in both the legislative and executive branches of government, thus indicating that they are ultimately accountable to an electorate, though the degree of accountability varies between countries as well as over time within each country.

Peru's democratic institutions provided for important constraints upon the president, but they depended upon the presence of a strong party system. While it is true that Fujimori contributed substantially to the weakening of the party system, it was already in decay by the time he took office in 1990. The judiciary was a weak constraint, not because of executive interference per se but rather because of its lack of independence historically and because it had been overwhelmed by the insurgency crisis of the 1980s. Peru's institutional context, both formal and informal, provided the chief executive with a great deal of policy leeway once the president was elected and his party controlled congress.

The operative institutional context within which Ecuador dealt with Peru was characterized by two key relationships; one between the legislative and executive branches and the other involving political elites, parties, and the electorate. In the legislative-executive relationship, congress held the upper hand. While the executive sought dramatic economic changes (liberal reforms), congress demonstrated repeatedly that it could and would block those policies. Congress's reach into the Supreme Court and tribunal courts prevented the judicial system from effectively playing a mediating or authoritative role in distinguishing between functions of the two branches of government. Only the military could rein in the congress, but it would have had to go beyond its formal institutional role, thereby increasing the potential costs to itself of becoming more politically active. In addition, its fear of becoming drawn into the day-to-day politics of governing generally kept it on the sidelines.

In terms of our hypotheses relating leader characteristics and institutional constraints (Tables 1.5 and 3.3), President Fujimori was an innovator and a risk taker who was faced with different levels of institutional constraints in 1995 and 1998. In 1995 the border skirmishes escalated into war at precisely the time of the campaign for the first elections under the new Constitution of 1993; neither the domestic nor the international legitimacy of the newly reformed democratic polity he had created could survive if he attempted to suspend or dramatically modify those constraints. Although Fujimori was a risk taker, he was not recklessly so. His risk aversion had to be high in 1995 because the credibility of his new political regime was at stake; consequently, we hypothesize that he reacted to events "by sticking to traditional policy positions" and that the hand of history was strong. By 1998, however, Peruvian new democratic institutions had achieved a level of legitimacy both domestically and internationally. With President Fujimori's party firmly in control of congress, he was in a position to act as an innovator and risk taker to push "new ideas while seeking to alter institutional constraints."

Ecuador's President Durán Ballén was a moderate innovator but very cautious. Thus we would expect him to be careful to take heed of the political winds before responding to a new situation, such as the outbreak of war. Presidents Bucaram and Mahuad were also risk takers and innovators, but Bucaram's recklessness virtually guaranteed that he would fall victim to the institutional constraints he faced. We would expect interim president Fabián Alarcón to play essentially an observer role, unwilling to push for any important policy changes even in the face of a deteriorating situation. President Mahuad, in contrast, was the Ecuadorean leader most likely to bet on a new policy if he believed that it would succeed and bring the opposition around to his side.

4 DOMESTIC POLITICS AND THE PUSH TOWARD WAR

This chapter analyzes the domestic political dynamics that produced war between Ecuador and Peru in 1995. Because the decision to respond to one of a series of minor border engagements by escalating to war was taken at the highest levels of government, we focus on the policy-making context in which the decision was made. While the broader context for the 1995 war includes both international and domestic factors, we argue that domestic considerations were determinant in producing the actual decision to fight. Therefore we focus our analysis largely on these factors.

In his analysis of interstate militarized disputes in Latin America, Mares has highlighted the key factors that determine whether policy makers would find it in their interest to engage in foreign military activities.[1] The basic argument is that leaders use foreign policy to provide collective and private goods to their domestic constituencies. The key question for leaders is whether the use of military force will benefit their constituencies at a cost that they are willing to pay and whether they can survive their displeasure if the costs are high.

The willingness of constituencies to pay costs varies with the value that they attach to the good in question, and their ability to constrain the leader varies with the institutional structure of accountability. The costs of using military force are influenced by the political-military strategy for the use of force, the strategic balance with the rival nation, and the characteristics of the military force used. A leader may choose to use force only when the costs produced by the combination of the political-military strategy chosen (S), the strategic balance (SB), and the characteristics of the force used (CF) are equal to or lower than the costs acceptable to the leader's constituency (CC), minus any slippage in accountability that is produced by the means of selecting leaders (A). Although force will not always be used when these conditions are met, force will certainly not be used in their absence. This dynamic may be expressed in the following equation:

$$S + SB + CF < CC - A \quad \text{may lead to the decision to use force}$$
$$S + SB + CF > CC - A \quad \text{no force will be used}$$

In the first section of this chapter we briefly recapitulate the positions of each country on the Amazon issue. The second section examines the factors affecting the costs associated with the potential use of force: S, SB, and CF. The third and fourth sections examine CC and A. The concluding section considers how and why a public within one democracy can in some circumstances be a stimulus for the use of force, even against another democracy.

Amazon Issue Foreign Policy

Ecuador's long-standing civilian and military leadership's concern for the Amazon issue revolves around its historical role in forging a national identity rather than serving economic or international political-strategic interests, which are minor at best.[2] Ecuador was created by bringing together vastly different provinces (the sierra and the coast), with very poor communications connecting them and with a large, historically unintegrated indigenous population. The glue that held the nation together internally consisted of caudillo, cacique, and familial clientelism in combination with the external threat posed by Peruvian claims to territory in the Amazon basin and southern coast. As in the case of Europe,[3] external threats played an important role in the creation of the Ecuadorean state and national identity. Occasional revolts against the traditional elites took the form of mass populism.[4] But the spasms of populism failed to create a new basis for Ecuador's national identity.[5]

Ecuadorean presidents since 1947 have wanted to settle the dispute with Peru, but not at the expense of the country's claim to sovereignty in the Amazon. The consistent foreign policy good was not settlement per se but rather a sovereign outlet to the Amazon. Achieving this goal would enable the president to recover some of the national pride and self-respect that most Ecuadoreans believed Peru had trampled on with its "aggression" in 1941, reinforcing national identity in the process. It thus constituted a "public good." Ecuadorean leaders did not pursue any private goods in this rivalry.

Peruvian presidents had a different foreign policy goal, reflecting that country's diplomatic and military advantages over Ecuador. Having won the 1941 war, Peru's leaders had to deliver its gains. Since virtually all Peruvians accepted the Rio Protocol, the 1942 treaty that formalized the peace and defined the border, defending it was a "public good." This meant that Peruvian leaders had to avoid getting caught in a diplomatic renegotiation that would give Ecuador sovereign access to the Amazon, in effect delivering a "public bad."

When President Fujimori (1990–2000) embarked on his neo-liberal development program for Peru in 1990, the building of economic linkages with neighboring countries became an important factor in Peruvian foreign policy. These potential economic links would constitute a public good to the degree that their benefits were widely distributed. If these economic benefits were limited to groups in the border area, however, they would be private goods. Fujimori certainly saw them in the broader context and hence as public goods.

Expected Costs of Using Force

Political-Military Strategy (S)

Ecuador first challenged the protocol settlement in the diplomatic arena, declaring it "inapplicable" in 1950 and "null" in 1960. In the mid-1970s the United States signaled that this strategy was appropriate when it said that Peru's position was too intransigent.[6] The democratic governments after 1979 followed their predecessors' leads. Ecuador appealed to the UN in 1980 and 1991 and to the OAS in 1980 and 1981 to take up the issue; in 1991 Ecuador suggested the pope as a possible mediator.

Ecuador's purely diplomatic strategy could not propel Peru to renegotiate. International actors were reluctant to reopen issues that had been legitimated in an international treaty, and Ecuador did not have the capability by itself to pressure Peru into discussing the issue. In the face of Peru's rejection of third-party involvement outside the parameters of the Rio Protocol, Ecuador's democratic governments followed a strategy of using military force to keep the issue alive and to induce third parties to intervene. Ecuador did not attempt to seize and control the disputed territory.

This strategy was predicated on not provoking Peru into escalating a conflict as long as Ecuador's armed forces were unable to hold out long enough for third parties to intervene diplomatically. The 1981 debacle, when Peru's forces quickly repulsed Ecuador's military incursions, demonstrated the continued weakness of the Ecuadorean armed forces. Over the next decade they developed their military capabilities and lulled the Peruvians with confidence-building measures among military personnel in the Cordillera del Cóndor zone.[7]

Ecuador's strategy after 1981 also took into account Peru's domestic institutional context. Ecuador's military command believed that the Peruvian military became demoralized and corrupted after a decade of fighting a civil war against both guerrillas and the drug trade (during which the institution was heavily criticized for human rights abuses and involvement in narcotics trafficking activities). Ecuadorean authorities also concluded that President Fujimori's interference with the military chain of command in order to assure personal loyalty further had damaged the institutional integrity of Peru's military.

Given these negative developments, Ecuador expected that Peru would be surprised by its military's defensive capabilities, waste significant resources in trying to overwhelm them, and be unable to adjust its local strategy before the costs of the war forced it to escalate or negotiate a cease-fire. Faced with significant losses in the Amazon, aware that Ecuador's navy had already left port, and observing the mobilized army in the south, Peru was expected to negotiate.[8]

The Ecuadoreans patiently waited for the right moment. In 1987 they discovered a new Peruvian outpost, Paquisha, in territory recognized as Ecuadorean during the post–Rio Protocol demarcation in the 1940s. Rather than denounce it at the time, they waited until 1991 to make it an issue.[9] During the 1991 controversy Ecuador did not back down; conflict was avoided by a gentlemen's agreement establishing a security zone and the mutual withdrawal of forces from two outposts. Neither side withdrew, however, producing a stalemate unchanged by minor Militarized Interstate Disputes (MIDs) in 1993 and 1994. These events suggest that Ecuador was ready to contest Peru militarily by 1991 but needed Peru to initiate the fighting.

Peru's political-military strategy was based on the sanctity of international treaties and did not change with the return of democracy in 1980. Peru argued that the Rio Protocol called for the four guarantors to resolve any disagreement within the parameters of the treaty. Under the terms of the treaty, Ecuador and Peru were to negotiate a separate instrument that would give Ecuador the right to transit Peruvian waters to the Amazon but not a sovereign outlet. Hence Peruvian leaders could ignore, if not explicitly reject, Ecuadorean calls for outside parties to intervene.

Within this Peruvian strategy, the use of military force was guided by two goals. The first was to keep Ecuador from effectively establishing outposts in remote disputed areas. The second was to resolve any military confrontation quickly, so as to avoid international pressure for a new basis for settling the dispute.

Diplomacy could produce economic benefits via increased cooperation. Both sides were aware that economic diplomacy might be a lever to induce the other side to make concessions. When the two countries were under military rule, Ecuador's attempt to tie the revival of the Andean Pact (an economic integration institution created in 1969) to a discussion of the territorial issue was quickly rejected by Peru. More than a decade after the return of democracy, when President Fujimori traveled to Ecuador offering economic cooperation as a means of developing a new bilateral relationship, Ecuador's presidents refused to accept any linkages with the Amazonian dispute.[10]

Strategic Balance (*SB*)

The balance of capabilities between Ecuador and Peru became more complex after 1980. As long as the dispute remained bilateral and the potential for escalation was great, the military balance appeared to favor Peru. Ecuadorean decision makers understood the fundamental disparity in military power.[11] Propelled by its defeat in the 1941 war, Ecuador's military began to support democratization to free itself from domestic politics and professionalize. Detailed analyses of military perceptions and justifications for supporting or threatening Ecuadorean democratic governments between 1948 and 1966 do not reveal intramilitary disagreements over how to proceed on the Amazonian issue as one of these factors.[12]

However, Ecuador's leaders did not believe that the balance of diplomatic capabilities favored Peru. First, Ecuador had demonstrated good faith in accepting the delimitation of over 95 percent of the border, as stipulated in the terms of the protocol. Second, Ecuador perceived that outside parties could recognize that the strong had trampled the weak in the 1941 conflict; with World War II over, however, the international community could remedy the injustice by insisting that Peru negotiate a relatively small (compared to what had been "lost") sovereign access to the Amazon. Although Peru repeatedly argued for the sanctity of international treaties, the declaration of inapplicability in 1950 did not question the treaty itself. Ecuador argued that the failure of the Rio Protocol to incorporate the real geographic situation made negotiations necessary.

Ecuadorean leaders recognized that the principle of righting an injustice would not by itself attract sufficient international attention, so they needed to keep the issue alive in order to persuade the international community that the dispute was not simply going to disappear and therefore that it needed to exert diplomatic pressure on Peru. The military skirmishes, renewed beginning in 1950, thus were directed at the international community, not Peru. The United States raised Ecuador's expectations that the international community might favor a "just" solution to the conflict in the mid-1970s when it critiqued Peru's position. The active role of the guarantors in terminating the 1981 miniwar, although carried out under the auspices of the OAS, indicated to Ecuador's leaders that they were on the right track if they could survive Peru's initial military response.

Nevertheless, Peru enjoyed diplomatic successes for almost fifty years: the OAS, the UN, and even the pope would not agree to mediate the dispute, because the Rio Protocol gave this task to its four guarantors. The 1981 experience convinced Peruvian leaders that they continued to have the military and diplomatic advantage, both because of Peru's quick military victory and because they concluded that the guarantors had interpreted their role on this occasion as simply helping to evacuate the Ecuadoreans safely and restoring the status quo ante.

Ecuador's armed forces' ability to resist Peruvian attacks in 1995 shifted the diplomatic balance, however. This time the guarantors worried that the conflict might escalate to large-scale war, especially as Peru kept committing more resources to the battleground in the Amazon. Durán Ballén, fearful that Peru could regroup and overwhelm Ecuador's positions, as in 1981, seized upon this new opportunity by convincing the international community that he had recognized the protocol and by committing the country to work with the guarantors for a resolution of the conflict.[13] Nevertheless, Peru maintained an advantage in that the mediators in the negotiations were the guarantors themselves, thus ensuring that the negotiations would not stray far from the 1942 Rio Protocol, which favored Peru.

Characteristics of Force Used (CF)

In the flare-up of the rivalry between the late 1970s and the 1990s, twelve minor MIDs (1977–1978, 1983, 1984, two in 1985, 1988, 1989, 1991, 1993,

1994, 1995, and 1998) and two major ones (1981 and 1995) occurred. From the 1950s to the 1980s Ecuador structured its penetration of disputed territory with only small units in isolated jungle outposts. In 1981 Ecuador constructed outposts on the eastern side of the mountain range, with poor lines of communication to Ecuador in the west. This was a quick and cheap incursion into disputed territory. Any direct confrontations with Peruvian troops would produce quick retreats or at worst a low number of casualties. Ecuador did not expect a strong response by Peru, especially not an invasion into Ecuador itself.

After the 1981 defeat, however, Ecuador's military leaders redesigned their approach to the use of force, with a shift in emphasis to its more efficient and successful deployment.[14] They chose terrain that would limit the maneuverability of hostile aircraft (with mountains at their back and steep mountains on each side, Peruvian aircraft could only come from one direction). The triple canopy jungle made it difficult to detect Ecuadorean defenses, thereby allowing soldiers to be hidden in trees with surface-to-air missiles. Planting cheap Chinese plastic antipersonnel mines made it difficult for Peruvian paratroopers to penetrate the area on foot. Weapons purchases seem to have been secret, ensuring that the Peruvian military would be unable to take effective countermeasures.[15] Effective lines of communication (a system of footpaths leading back to Ecuadorean base camps and villages) were developed. The Ecuadoreans also contracted Israeli and Chilean intelligence and communication experts to create a system to intercept Peruvian communications. Finally, they prepared national defenses in case of escalation, including getting the navy out of port quickly.[16]

In Peru's case, between 1981 and 1995 the force characteristics employed for confronting Ecuador did not change. Peruvian military leaders continued to believe that they retained military superiority, with modern fighter-bombers, attack helicopters, and well-trained paratroopers able to inflict another quick and inexpensive defeat on any Ecuadorean incursions. They also felt that Peru could overcome any initial losses, if they occurred, due to high levels of equipment in reserve as a result of a military buildup in the 1970s in preparation for a possible war with Chile.[17]

Because of the military power disparity and Ecuador's exposed positions, in 1981 Peru's response had produced a quick and cheap victory. At that time Peru had enough confidence in its military superiority to threaten an invasion of Ecuador as well. But in 1995 Peru's miscalculation of its opponent's improved military capacity and the difficulty its armed forces had in responding effectively to it contributed to the leaders' hesitation to attempt to pursue victory this time. However belatedly, Peru's recognition of these new *CF* realities indicates that it understood that the military costs of large-scale war were now significantly higher than before.

Cost Summary

MIDs were low-cost options for Peru as long as it could overwhelm Ecuadorean outposts easily. The 1941 war had been cheap for Peru, as had the 1981 conflict. The potential military and diplomatic costs of renewed conflict had increased greatly by 1995, however, and even more so to the degree that the international community became involved. For Ecuador, MIDs were low-cost options as well, as long as Peru did not escalate. Ecuador's defeat in 1981 had been expensive and had not advanced its political-military strategy. By 1991 a defensive war looked possible, though still costly. From Ecuador's perspective, if members of the international community intervened quickly as a result of war, they would likely pressure Peru to make some concessions to resolve the dispute.

Constituency Cost (CC) Acceptability for Ecuador and Peru

Ecuador

Democratic politicians after 1979 could draw on past experience to evaluate what their constituencies wanted and what costs they were inclined to accept. Ecuador remained democratic throughout the 1948–1960 period, during which its strategy to challenge the 1941 settlement was conceived and implemented. The distinct constituencies of the three presidents elected over these years suggest that a foreign policy demanding recognition of Ecuador's sovereign access to the Amazon represented a broad national consensus.

President Galo Plaza Lasso (1948–1952) was a moderate, with good relations with the United States. José María Velasco Ibarra (1952–1956) won the presidency behind a conservative and populist alliance, as did Camilo Ponce Enríquez (1956–1960) and Velasco Ibarra (1960–1963) a second time. President Plaza had little internal opposition when he declared the protocol "inapplicable." President Velasco Ibarra had just been elected, with more votes than the combined total of all his opponents, when he declared the treaty "null" in 1960.

After the country's poor performance on the battlefield and in the OAS during the miniwar of 1981, Ecuador's foreign ministry undertook a national opinion survey on the issue to update the government's evaluation of national sentiment. The diplomatic corps perceived Ecuador's strategy as fundamentally flawed because it contested the principle of the sanctity of treaties and sacrificed national development to a vague territorial issue. However, the opinion poll confirmed the popularity of the strategy of nullification and sovereign access.[18] In 1983 the Ecuadorean congress reiterated the country's claim that the protocol was null and void.[19]

Another poll carried out in 1992 provided further evidence of Ecuadorean feeling on the issue. The overwhelming majority of Ecuadoreans believed that the border issue obstructed development (79 percent yes, 15 percent no), and a majority believed the country should have free trade with Peru (55 percent yes, 39 percent no). These data indicate that Ecuadoreans were well aware of the economic costs resulting from the dispute. Nevertheless, 49 percent believed Peru to be an "enemy" country, compared with only 39 percent who perceived it to be friendly, making it clear that Ecuadoreans expected Peru to make the concession that would improve relations.[20]

Ecuadoreans were not ignorant of the costs of continuing the conflict: 53 percent of Quiteños and 38.5 percent of Guayaquil respondents in a 1996 poll believed that Ecuador was more affected economically by the war than was Peru, and an overwhelming 80 percent believed that armed confrontations would recur. In addition, while most believed that Ecuador had "won" in 1995 (55.0 percent in Quito and 74.3 percent in Guayaquil), respondents were far more pessimistic regarding the country's ability to prevail in a new confrontation (39.5 percent in Quito and 52.0 percent in Guayaquil).[21]

However, Ecuadorean presidents have military as well as civilian constituents. The military, while not interested in governing after 1976, was concerned about the territorial issue. Civilians wanted the military to professionalize, both in order to implement the country's political-military strategy on the border and to ensure the continuation of democracy. Since democracy's return in Ecuador in 1979, four elected presidents had overseen the expansion of their military institution's capability. As a result, military capacity increased substantially under democratic auspices through increased expenditures, even as the military's share of Gross National Product (GNP) declined.[22] Ecuador's military learned the lessons of its embarrassing defeat in 1981 and looked for the government to support its efforts to reverse future results.[23] The Ecuadorean military was not anxious for a large-scale war, both because the outcome was uncertain at best and because the economic costs to the country would almost certainly be devastating.

Peru

Since returning to democracy in 1980, Peru had three elected presidents during the time frame under consideration: Fernando Belaúnde Terry (1980–1985), Alan García Pérez (1985–1990), and Alberto Fujimori Fujimori (1990–2000). Fujimori was also the head of state during the authoritarian interlude between the dissolution of congress in April 1992 and the plebiscite in November 1993 confirming the new constitution. While each leader had distinct constituencies, none of the coalitions expressed any interest in resolving the dispute with Ecuador by renegotiating the Rio Protocol.

Belaúnde, elected by a broad national coalition, received 45 percent of the presidential vote, compared to the runner-up with 27 percent. This was the first elected government after the transition to democracy, so the military represented an indirect constituency as well. Belaúnde's electoral coalition would collapse in protest over economic and social policy as well as corruption.[24] Yet there was no pressure to change policy toward Ecuador. The military leaders wanted a convincing response in 1981 to deter future incursions by its northern neighbor, and Belaúnde worked closely with them.[25] Lending support to this approach were strong expressions of public sentiment in favor of expelling Ecuador from the area.[26]

García's populist coalition of center-left parties controlled congress, with 105 of the 180 deputies and half of the 60 senators (the other half were distributed among fifteen other parties).[27] The leading constituencies in this multiclass alliance were domestically oriented business, the middle class, and the urban working class. The promise of resources distributed by the state and a nationalist ideology brought the alliance together. An economic crisis after 1987, largely the result of the populist program itself, split the alliance and led to García's dramatic fall in public approval ratings from over 90 percent at the beginning of his term to just over 10 percent at the end.[28] Judging by the absence of the dispute in academic analyses on the period, the continued conflict with Ecuador seems not to have been a factor in García's rise and fall or in his own views of the pressing issues facing the country.

Fujimori's electoral coalition was not party based and did not control congress from 1990 to 1992 but did have a majority in the Constituent Congress (which wrote the Constitution of 1993) from 1992 to 1995 and the regular congress after the 1995 elections.[29] He won the 1990 elections in the second round with 62.4 percent of the vote, despite being outspent by runner-up Mario Vargas Llosa by 60 to 1.[30] The success of the Fujimori government's market liberalization reforms in controlling inflation and promoting growth, as well as its achievements in combating guerrilla violence, made the president very popular both in Peru and abroad.[31] Because working toward a better economic relationship with Ecuador was important for the administration's development program, Fujimori initiated efforts to improve relations within the context of the 1942 protocol. Pre-1995 efforts were limited to economic incentives, however, and generated little opposition.

In general Peruvians had a positive disposition toward Ecuador. In a poll conducted in January 1994, a year before the 1995 war, 63 percent of respondents perceived Ecuador as a "friendly country" and only 23 percent as an "enemy." (Many Peruvians, however, do refer derogatorily to Ecuadoreans as *monos* [monkeys]. Combined with the territorial issue, this attitude may help explain why 49 percent of Ecuadoreans saw Peru as an enemy.) In an April 1994 poll, 41 percent of Peruvians believed that "no problem" existed between the two countries because the protocol had resolved it. Of those who saw a problem, more than half believed that the guarantor countries of the protocol should arbitrate it. Fully 73 percent believed that demarcation should proceed along the lines of the protocol.[32]

Peruvians believed that Ecuador had been progressively intruding on Peruvian territory: 65 percent thought that the troops had been there before 1994, and another 16 percent that they had arrived in 1994. During the 1995 war, two opinion polls in February found overwhelming support for the actions of the armed forces (86.5 percent and 88.4 percent), but a bare majority approved the behavior of the guarantor countries (54 percent and 57 percent). Fujimori's military response during the war was supported by 59.2 percent; his chief rival, Javier Pérez de Cuellar, who advocated a more forceful response,[33] received the approval of 46.6 percent. At the same time, many Peruvians were aware that this conflict would not be a repeat of the miniwar of 1981; just 34 percent thought that it would take one to two months to expel the invaders, while another 29 percent believed that it would take longer. The war was important enough to Peruvians that 55 percent believed that presidential elections should be postponed if the war continued.[34] In short, Peruvians supported the war effort, although they desired peace.

Leaders' Accountability to Constituencies (A) in Ecuador and Peru

Ecuador

Ecuadorean civil society is well organized and willing to engage in pressure group activity independently from its representatives in congress. A group of indigenous communities that had begun to organize in the 1980s, the Confederación de Nacionalidades Indígenas del Ecuador (CONAIE, Confederation of Indigenous Nationalities of Ecuador), went on a national strike in 1991 to call attention to their economic and political plight. By 1996 they had become an important political party, Pachacutik. Student groups, business associations, and unions went on a national strike in 1997 to protest against President Bucaram's administration. Mass demonstrations in the main plaza of Ecuador's capital, Quito, became a common occurrence.

Because of popular support for a sovereign access to the Amazon, presidents were unwilling to follow the foreign ministry's perception that the issue needed to be resolved under the available means (the Rio Protocol). Something had to change in a positive direction for Ecuador before the public could perceive that a resolution with sovereign access was acceptable, and no president could bring about those changes by himself.

As noted in Chapter 3, the armed forces were the most respected institution in the country during this period, far outdistancing either congress or the presidency.[35] Popular support provided a major basis for military influence in politics, but the historical record does not suggest that presidents perceived that the military was keeping them on a tight leash with respect to the territorial dispute.

We do not know the details of the decision to move Ecuadorean troops into the disputed territory; but it was not a secret move by the military, because President Jaime Roldós (1979–1981) visited the outposts in August 1980.[36] Once the fighting began in 1981, it is possible that the civilian government had little control over military operations.[37] President Roldós made the decision to ask the OAS and the "four friendly countries" to mediate without consulting the military, however, whose leadership did not seek to overturn the request.[38] Although the military may not have entirely agreed with President Sixto Durán Ballén (1992–1996) in his 1995 decision once again to accept the validity of the Rio Protocol as the best way to terminate the outbreak of war with Peru, it did not oppose the decision.[39]

It was clear that the Ecuadorean military could weigh into a dispute and affect the outcome, so a president had to consider its views. Because the military was not interested in governing, however, the accountability of presidents to that institution and its leadership was indirect and depended more on the military's influence with civil society. Politicians were failing to resolve the various political and economic crises that confront Ecuador, making public clamor for a more direct military role in governing a distinct possibility.[40]

Durán Ballén did not want war but could not unilaterally withdraw in the face of initial Ecuadorean military success in resisting Peruvian efforts to dislodge its forces from Tiwintza, not only because the military opposed retreat but also because the public would not have accepted it. His government had been weakened by the economic crisis and a failed referendum to force economic policy changes blocked by congress. The war produced a "rally around the flag" effect, however, and lifted his approval ratings.[41] Fearing the economic and military consequences of further military escalation, Durán Ballén quickly improvised. He made a deliberately ambiguous reference to the guarantor countries to mediate the dispute, asking for an emergency meeting of the Rio Protocol guarantors to "inform them of the most recent border incidents . . . [and to ask] these countries for help in resolving them."[42] While the guarantors and the Peruvians interpreted this request as recognition of the protocol, in Ecuador the ambiguity resulted in a perception that the mediation request included reexamining the "illegitimate" treaty that was the underlying basis of the dispute.[43]

Yet the willingness of the guarantor countries to mediate did not mean that Durán Ballén would call off the troops. Seeking to recover the initiative, Fujimori and the Peruvian armed forces militarily engaged the Ecuadoreans despite declarations of cease-fire; the Ecuadoreans retaliated and stood their ground in Tiwintza. Continuing to fight perceived Peruvian "aggression" was too important for the public, and the military desperately desired to preserve its localized success. Consequently, Durán Ballén continued to pay the costs of war until Fujimori accepted the military status quo.

Peru

President Fujimori's accountability to his constituency in early 1995 was probably at its high point; as noted in Chapter 3, the very legitimacy of his new project for Peru was at stake, as embodied in the 1992 coup, the 1993 Constitution, and the April 1995 election run-up in which he was a candidate for reelection. Fujimori did not try to govern without popular support; not only did his coup in 1992 have overwhelming approval and his 1993 Constitution pass in a referendum, but when scandal caused that support to collapse in 2000 he left office rather than attempt to extend his rule by violence. Consequently, the decisions he made during the war implied great domestic political risks for him. Risk acceptance was not his only personality trait at work during this time; his personal authoritarian bent demonstrated his commitment to impose high costs on those who opposed his major projects.

Fujimori seized upon the 1991 flare-up to push for a definitive settlement, although he was criticized in congress for not dealing harshly with Ecuador.[44] In 1991, 1992, and 1993 he offered Ecuador a package linking economic development projects, a free port on the Amazon, and reciprocal security measures and arms limitations along the border in exchange for a border demarcation associated with the Rio Protocol. His 1991 trip to Ecuador to offer details on the proposals represented the first by a Peruvian president; he would eventually go three times.

At the same time when Fujimori was extending the olive branch (on Peruvian terms), he demonstrated his unwillingness to compromise on fundamental points. In early 1991 Ecuador asked privately that Peru abandon the disputed outpost. Peru initially responded with threats, then accepted a gentlemen's agreement on mutual withdrawal of forces, which soon collapsed after the Peruvian military expressed its vehement opposition,[45] setting the stage for war in 1995.

We can gauge some of Fujimori's accountability to the people and congress at work during the 1995 war, as it coincided with the presidential campaign. Fujimori's major opponent, former UN secretary general Javier Pérez de Cuellar, and his military advisors publicly wondered about Fujimori's ability to defend Peruvian interests and called for more severe action against Ecuador. In response, Fujimori claimed that his conciliatory policies had been designed to deceive Ecuador and increased Peru's efforts to win on the battlefield.[46] Fujimori was not planning to cede sovereign access to the Amazon to Ecuador, and regional economic development fit in with his neo-liberal outlook (he sought similar programs with Peru's major rival, Chile). Hence one should take this claim to be defensive campaign rhetoric. In March he offered Ecuador the carrot of a possible free trade zone in the Amazon but retracted it as a result of congressional opposition.[47]

Fujimori won in a landslide in April, although his victory probably had more to do with his success in bringing down inflation, stimulating economic growth, and curbing the Shining Path guerrillas as well as a deceptive public relations campaign in the media than with the war with Ecuador.

Assessment of Cost Acceptance and Leadership Accountability

Constituencies in both countries were knowledgeable about the dispute and favored a resolution, but each expected the other side to make greater concessions. Both countries' sets of constituencies repeatedly demonstrated a willingness to accept the costs of war if necessary to defend their countries' interests in the Amazon. As long as Peru refused to recognize the legitimacy of their claims for renegotiation, Ecuadoreans supported the militarized bargaining strategy even at the expense of the economic benefits they expected from better relations with Peru. For their part, Peruvians supported using military force to defend the Rio Protocol.

Redemocratization did not imply greater accountability of elected heads of state to either the legislature or the electorate. In Ecuador, this slippage after election was largely the result of the prohibition on reelection and the weakness of the party system, especially in congress. In Peru, presidents were theoretically more accountable because they could compete for presidential office again after sitting out at least one term. President Belaúnde did not harbor expectations of winning another term, however, because he was already in his seventies. President García felt little accountability to the electorate, apparently miscalculating that the APRA party could survive voter disapproval of his actions as head of state. President Fujimori, in contrast, was very accountable in 1995, not only because the war broke out during the presidential election campaign but also because this was the first election under the new constitution and was supposed to complete Peru's return to democracy after the 1992 coup. Fujimori needed to win this election with a minimum of fraud, so his accountability to the electorate was very high during the war.

Conclusion

The 1995 war was popular in both Ecuador and Peru. Each side would have preferred to attain its goals without the use of force, but both sides were willing to use military action rather than simply accept the other country's definition of the border. Policy makers understood these sentiments and thus were prepared to raise the stakes in 1995 when a new opportunity emerged for Ecuador to press its claims.

Ecuador's population disputed the terms of the treaty ending the 1941 war. Ecuador's leaders consistently sought the means to challenge the status quo, not as a diversionary tactic but because their constituencies wanted a favorable resolution of the Amazon issue. The geographic anomaly in the Rio Protocol provided Ecuador with an opportunity to devise a political-military strategy to pursue an outlet to the Amazon. Initially, however, it had neither the diplomatic nor the military ability to persuade third parties to pressure Peru into negotiating a new settlement. Changes in the characteristics of the force used, as well as in planning its use, were implemented across four different presidencies. These changes produced a shift in the strategic balance by 1995 at a cost acceptable to the majority of the population and to the military as well. Seizing the diplomatic initiative as peacemaker for the first time ever in its conflict, Ecuador enabled the guarantor countries to play an active role within the Rio Protocol for the first time in over forty years.

Peru was the defender of the post-1941 status quo that granted Ecuador no sovereign access to the Amazon. Whereas Ecuador had an appreciation for the complexity of militarized bargaining, Peru did not. Fujimori's economic carrots were not linked to military policies that could have deterred Ecuador's strategy. All Peruvian presidents adopted a straightforward political-military strategy for defending the protocol. As a defender of an internationally recognized status quo, Peru refused to reopen the question. Coercive diplomacy was promised at the local level to dissuade incursions, and another blitzkrieg into southern Ecuador was threatened if the first approach failed. While these military policies were an appropriate response in 1941 and 1981, Peru's strategy became vulnerable to outside party influence when its military capability diminished.

Because neither the Peruvian military nor Fujimori had a dynamic sense of the strategic balance and the population would not accept defeat, Fujimori found himself forced to escalate the fighting. Not only did it fail to completely dislodge the Ecuadoreans, but in the eyes of many of its traditional international supporters Peru transformed itself from a defender of the status quo into a threat to regional peace.

5 THE DOMESTIC BASES FOR RESOLUTION

The 1995 war changed the international, bilateral, and domestic contexts in which this border dispute evolved. This chapter analyzes the domestic political dynamics among civilian politicians, diplomats, and military officers in Ecuador and Peru after 1995 that permitted adjustments in the negotiating positions of the two sides, thereby making resolution a realistic possibility. In Chapter 6 we examine the international changes produced by the war which helped Ecuadorean and Peruvian leaders actually achieve resolution.

The analysis does not focus on which side was "right" or make a judgment on the merits of the cases presented by the two sides. Over 165 years of tension and conflict regarding the border issue clearly demonstrated that each side perceived its position to be sufficiently correct and honorable to defend with war. Rather, we focus on the way in which the result of the 1995 war changed the incentives of key players in the domestic institutional context in ways that made a settlement possible. Most of the important shifts in incentives came on the Ecuadorean side, both because just about everyone in Ecuador and elsewhere perceived Ecuador as the winner of this war (see Chapter 2) and because Ecuador continued to find itself in the weaker diplomatic position.

Ecuador's diplomatic weakness on the border issue stemmed mostly from its efforts over decades to seek revisions of a treaty settlement that both parties had signed, which was long recognized by the inter-

American community and was considered valid under international law. In addition, due to its smaller size and greater susceptibility to economic problems,[1] Ecuador stood to benefit more than Peru from the increased economic interactions that resolution of the conflict was likely to produce. Therefore, the country had a greater incentive to shift its position on economic grounds as well.

These weaknesses existed before the 1995 war. So a key question is why Ecuador did not shift its negotiating position earlier. The answer is primarily that the dispute involved Ecuadoreans' underlying sense of identity as a nation, transcending the actual physical territory claimed (see the discussion in Chapter 4). Any solution to the dispute therefore had to provide Ecuador with both a concrete basis for its claim to be an Amazonian nation and a conviction that Peru was dealing with it as an equal.

In addition, to be viable, any agreement also had to be in the interest of the political leadership itself. This chapter analyzes the conditions under which the Ecuadorean and Peruvian civilian leaders, diplomats, and armed forces found it in their respective interests to resolve the border dispute by focusing on strategic interactions among the relevant institutional actors. We do not privilege either democratic processes or economic interdependence per se as determining conflict resolution, in spite of the abundant literature positing such a relationship.[2] Rather, we examine the specific interests of each set of domestic actors within the Ecuadorean and Peruvian institutional contexts (analyzed in Chapter 3) to indicate how continuation of the conflict and specific resolutions of it affected those interests.

The first section of this chapter examines public opinion on relations between the two countries at the time of the war and during the settlement negotiations. The subsequent three sections focus on the interests of diplomats, the military, and politicians as they relate to the border issue.

Public Opinion and the Border Dispute in Ecuador and Peru

Ecuador

Public opinion on relations with Peru and the border issue was ambivalent, and its implications for policy were not straightforward. Ecuadorean public opinion showed a distrust of Peru, a split on the traditional nationalist stance regarding the border issue, and a concern over the possibility of war but, at the same time, a willingness to engage in war rather than give in to Peruvian demands. The dramatic outpouring of support for President Durán Ballén when the 1995 war broke out led many analysts to conclude that popular sentiment on the need for Ecuador to gain sovereign access to the Amazon River made it impossible for Ecuadorean leaders to resolve the issue. Nevertheless, the evidence also suggested that, if a settlement package was properly structured, broad popular support for a comprehensive resolution could be forthcoming.

Following the war, Ecuadorean public opinion seemed to demand greater concessions in order to settle. In early 1995 (during the war), an opinion poll asked, "Do you believe that it is possible for Ecuador to recover all the territory lost in 1941?" In Quito 27.5 percent responded yes to the question, while in Guayaquil the figure was 32.8 percent. A year later, in 1996, when provided with the statement "There are people who say that Ecuador should recover the territory which it lost in the 1941 war and that it should be done no matter what the cost," 44.3 percent of Quiteños and 56.5 percent of Guayaquileños agreed! On the issue of sovereign access to the Amazon, a significant majority in 1995 believed that it was possible despite the conflict (75.3 percent in Quito, 83.5 percent in Guayaquil; see Appendix C).

However, polling questions that examine the border issue in isolation do not provide a sense of the importance of this issue in relation to others. A poll carried out in June 1996, just before the final round of the presidential election that year, asked potential voters in both Quito and Guayaquil if the candidate for whom they planned to vote would do better than his competitor on seven issues. One of these was negotiating with Peru, a particularly timely concern because the two countries had been engaged, along with the guarantor states, in working for the border dispute resolution for over a year.

These poll results indicated that, for voters likely to support Bucaram, the difference between the two candidates on negotiating with Peru was less important than the issues of controlling inflation, stabilizing the economy, decreasing poverty, combating corruption, and diminishing the number of strikes. For likely Jaime Nebot voters, however, negotiating with Peru and stabilizing the economy were the two issues on which they saw the greatest difference between the two candidates. In the 1996 elections, Abdalá Bucaram won in a landslide, beating Nebot in every province except Guayas, home to both candidates.

The polling responses cited above (elaborated further in Table 5.1), along with Bucaram's overwhelming victory, indicate that the border dispute with Peru was an important issue but not the defining one in Ecuadorean politics. This finding suggests that members of the public would be prepared to accept a comprehensive settlement with Peru if

Table 5.1. Performance Expectations among Likely Voters (%)

Among Those Likely to Vote for Bucaram					
Question #18	Jaime Nebot	Abdalá Bucaram	Both	Neither	Does Not Know/No Response
There would be fewer poor	3.2	70.3	6.0	17.6	2.9
Prices would increase less	2.2	79.7	4.1	11.1	2.9
Would negotiate better with Peru	20.4	55.7	8.5	7.6	7.7
Economy would be more stable	7.3	74.9	4.2	9.3	4.3
There would be less corruption	5.3	71.0	8.7	11.7	3.3
There would be fewer strikes and work stoppages	6.1	68.1	7.3	11.2	7.3
Public employees would be let go	34.5	43.3	8.1	5.9	8.4
Among Those Likely to Vote for Nebot					
Question #18	Jaime Nebot	Abdalá Bucaram	Both	Neither	Does Not Know/No Response
There would be fewer poor	63.4	5.1	3.9	24.2	3.4
Prices would increase less	65.3	8.6	6.1	15.2	4.8
Would negotiate better with Peru	86.6	4.2	1.2	3.9	3.9
Economy would be more stable	84.3	4.3	2.0	5.6	3.7
There would be less corruption	74.9	5.3	4.0	13.4	2.4
There would be fewer strikes and work stoppages	70.8	4.7	6.2	13.4	4.9
Public employees would be let go	47.0	30.0	9.2	6.3	7.4

Source: Perfiles de Opinión (Quito: Perfiles de Opinión, Cía. Ltda.), No. 22, June 1996, 49–50.

they perceived that supporting the status quo on the border would undermine their president's ability to achieve what they considered to be more important goals.

Nevertheless, the border issue could still contribute to a president's political troubles. Bucaram became the first Ecuadorean president to travel to Peru, for which some groups criticized him. In a speech before the Peruvian congress, however, when he called for both sides to apologize, the uproar at home was nearly unanimous. Ecuadoreans felt that they had no need to apologize, given their view that it was Peruvian aggression which had produced the problem in the first place. This faux pas contributed to Bucaram's impeachment a month later.[3]

Peru

As noted in Chapter 4, Peruvian perceptions of Ecuador before the 1995 war were more positive than the views of their Ecuadorean counterparts in regard to Peru. This was probably because a large plurality believed that the border issue had been resolved by the 1942 Rio Protocol and most of the rest believed that demarcation could only proceed according to its provisions. Like the Ecuadoreans, however, Peruvians rallied around their armed forces once war broke out in January 1995.

Nevertheless, Peruvian opinion polarized around the manner in which the conflict could be resolved. A small majority in the two February 1995 polls (59 percent and 54 percent) supported mutual concessions, but 27 percent in one poll and 40 percent in the other were opposed to any concessions at all. At the end of the war, 61 percent believed that no one had won and 25 percent that Peru had won. Another war was seen as very likely by 27 percent and likely by 40 percent. Only 13 percent thought it highly unlikely (*nada probable*).[4]

In short, Peruvians generally supported the war effort, although they desired peace. A significant minority were unwilling to make any concessions to Ecuador for peace, but the majority favored accommodation as long as it was within the parameters of the Rio Protocol.

Bureaucratic Politics and the Diplomatic Corps

Whether bureaucracies have an important impact on policy depends on how the decision-making unit is structured as well as the degree of the bureaucratic leadership's independence from the chief executive.[5] Although one might assume that a foreign ministry would have a tremendous impact on all foreign policies, policy making can be structured in such a way as to exclude or minimize the foreign ministry's influence. In addition, if the president selects ministers because they represent certain political forces, the bureaucracy will be more likely to influence the policy process than if the ministers were selected for their personal ties to the chief executive.

Ecuador

Members of the diplomatic corps in Ecuador were very professional and capable but ultimately captive to the political leadership. They had no independent bargaining power to use at home; the international reputation of Ecuador was not a powerful tool in domestic politics. As agents of the president, and subject to censure by the congress simply for policy disagreements, career diplomats in the foreign ministry would gain no advantage by blazing an independent path around new options to resolve the border dispute with Peru. Diplomats would play a role if Ecuador decided to accept nonsovereign access to the Amazon, because they would have to package the deal with Peru in such a way as to make it salable at home.

Mares interviewed middle- and upper-level ministry officials (in confidence) and interacted with them during a two-week course at the Diplomatic Academy of the Foreign Ministry of Ecuador a few months after the war. These interactions made it quite clear that many career diplomats were aware of the limitations of Ecuador's position on the protocol and were willing to resolve this issue in line with its provisions in order to move on to better economic relations with Peru.

It was not a generational issue. One former minister commented that declaring the Rio Protocol "null" was a terrible mistake for Ecuador because it put the country in the position of unilaterally opposing an international treaty whose legitimacy had been recognized by all other relevant parties. During the administration of Rodrigo Borja Cevallos (1988–1992), in fact, the foreign ministry had been quite creative in putting together possible packages for negotiations. In 1991 foreign minister Diego Cordovez made remarks which suggested that Ecuador might moderate its traditional demands and proposed increasing economic relations as a means to lay the groundwork for a future settlement.[6]

The Ecuadorean foreign ministry was ready to move ahead; the diplomats just needed the politicians to indicate that a solution without sovereign access to the Amazon, which was the fundamental conclusion of the Rio Protocol, was acceptable. Such signals came quickly: Bucaram retained foreign minister Leoro Franco in his post and consulted closely with him. Bucaram's decision to become the first Ecuadorean president to visit Peru and his courteous interactions with his hosts gave a strong signal of support for negotiations. Even after Bucaram's tumultuous ouster and the naming of a new foreign minister, José Ayala Lasso, the foreign ministry retained the lead in the negotiations.

Under pressure on domestic issues as the economy began to crash and the social movements that overthrew Bucaram demanded political reforms, his successor, interim president Fabián Alarcón, could not devote the necessary time and attention to molding Ecuador's negotiating position. Thus he accepted the foreign ministry's strategy to inform and consult with the relevant political, economic, social, and media forces at home in order to build a domestic consensus on the general negotiating strategy. Francisco Carrión Mena's detailed analysis of these meetings never refers to Alarcón making any contacts or saying anything about the peace process to these groups; instead, all of the public and private meetings were managed by the ministry.[7]

It is not that President Alarcón did not care about the negotiations or their potential to create domestic problems. After almost a year in office and with the negotiations moving into more complicated terrain, he began to worry that society might not be ready to accept the terms the ministry was indicating as likely. In January 1998 Alarcón publicly suggested that a national plebiscite might be necessary before the negotiations proceeded further. The ministry held an internal discussion and communicated its fear that a plebiscite could derail the negotiations by politicizing them and by undercutting the guarantor countries. Alarcón insisted on consulting with the three Latin American presidents among the guarantors. When they expressed concerns paralleling those of the foreign ministry, however, Alarcón publicly backed away from his proposal, thereby ensuring that the ministry would remain in charge.[8]

Peru

The diplomatic corps in Peru was a highly professional body, dating back to the 1850s and trained as a career service for decades in a highly regarded diplomatic academy.[9] Early in his administration, however, President Fujimori sought to impose his own stamp on the foreign ministry by naming noncareer ministers and by using his post-April 1992 *autogolpe* decree powers in December 1992 to purge the diplomatic corps of 117 of its most experienced professionals and end the long-standing merit-based promotion system.[10] With these initiatives, Fujimori gained virtually complete control over the administration of Peru's foreign affairs well before the outbreak of hostilities with Ecuador in 1995.

Due to the severity of internal political and economic crises plaguing Peru in the early 1990s, including the generalization of violence by Shining Path guerrillas, hyperinflation, and the virtual collapse of the government bureaucracy, Fujimori worked to reduce tensions with Peru's neighbors in order to concentrate on these domestic problems.[11] He engaged in his own personal diplomacy with Ecuador, Chile, and Bolivia to accomplish this goal. When problems arose in this pursuit, however, as in the negative reactions by congress and the military to his initiatives in Ecuador, for example (see the discussion in Chapter 4), he responded by sacking his foreign minister, Carlos Torres y Torres Lara.[12] These diplomatic forays did permit Fujimori to redirect military forces to combat the threat posed by Shining Path. Ecuador's lack of formal response to his diplomatic proposals in 1991 and early 1992 served only to postpone the day of reckoning with Peru's northern neighbor, however, while the Fujimori government's attention was focused on overcoming major domestic problems.

Once the conflict with Ecuador was moving to serious diplomatic negotiations with the help of the guarantors, Fujimori named a hard-liner, Eduardo Ferrero Costa, as foreign minister in July 1997. He represented the dominant popular sentiment (fanned by an increasingly government-manipulated Peruvian media) that the only acceptable resolution of the dispute was the application of the Rio Protocol with no concessions whatsoever.[13] This rigid stance represented the Peruvian government's public position until very shortly before a definitive resolution was achieved in October 1998. It helps to explain why violent protests broke out in Peru when the final terms were made public and why Ferrero Costa had resigned shortly before, after failing to prevent the symbolic territorial concession that enabled Ecuador to accept the final settlement.[14]

The Armed Forces

It is commonplace in the study of Latin American politics to perceive the military as xenophobic, antidemocratic, and focused on short-term organizational needs (such as weapons, personnel, and pensions). As a result, cooperative regional relations seem to depend upon civilian control over the military; Peruvian behavior during the negotiations was in line with these expectations, largely because Fujimori had transformed Peruvian civil-military relations by the time of the 1995 war from a relationship of parallel spheres of action to one of civilian dominance. As the complexities of the Ecuadorean case also illustrate, however, such a straightforward view of the military institutions' posture is incorrect.

Ecuador

The civil-military relationship in Ecuador retained the character of parallel spheres of influence before, during, and after the war. This is not to argue, however, that civilians were unable to set policy: President Galo Plaza provoked no institutional opposition when he closed the Military Academy from 1953 to 1956 after some officers grumbled that his budgets would keep Ecuador weak vis-à-vis Peru. Although the military may not have entirely agreed with the civilian presidents' decisions on how to terminate the military conflicts in 1981 and 1995, it did not resort to institutional actions to oppose them.

Given this institutional dynamic of far less than full civilian control in Ecuador, one might reasonably expect the country's military to constitute a major obstacle to accepting a treaty that did not provide for sovereign Ecuadorean access to the Amazon. Our conclusion, based on extensive interviews with Ecuadorean officers and a review of Ecuadorean military literature and speeches, is quite different.

The dramatic defeat at the hands of Peru in 1941, and again in 1981 when Ecuadorean forces quickly withdrew in the face of Peruvian military superiority, propelled the Ecuadorean military to professionalize in order to have more success in carrying out its tasks, defined as both defending sovereignty and contributing to the economic and social development of the country. The military leadership recognized that professionalization required turning over the day-to-day operations of government to civilians. While the military did accept democracy, officers did not see themselves as precluded from playing a role internally if civilian government threatened or produced political instability. Consequently, the military intervened briefly in 1964–1966 and again for a longer period as an institutionalized military regime between 1973 and 1979. This was the one significant if temporary variation in the parallel spheres of influence dynamic that generally prevailed.[15]

The Ecuadorean military endorsed a return to democracy after 1976 in order to focus again on professionalizing and modernizing its force structure.[16] Civilians also worked to achieve military professionalization as a process that would increase the likelihood of democratic continuity. Since the return of democracy to Ecuador in 1979, four presidents have overseen the expansion of military capability and institutional capacity, even though the military's share of GNP declined during these years of civilian political control.[17]

Continued clashes with Peru on the disputed border further stimulated professionalization efforts in the military. Interviews by Mares as well as published accounts in Ecuador indicated that Ecuador's military leaders learned the lessons of their embarrassing defeat in 1981. The military undertook a sophisticated analysis of not just the military factors but also the domestic situation in Peru and the likelihood of regional diplomatic mediation and turned the tables on Peru in 1995.[18]

What were the implications of the military's professionalization for the process by which the border issue with Peru was negotiated? Our analysis points to two major factors that impacted negotiations with

Peru. First, the military could not play the leading role in resolving the dispute. The military could not make the jump from a focus on territorial defense to advocacy of a comprehensive national interest. For the military to declare unilaterally that it accepted the Peruvian decision on border demarcation would call into question its military budget and prerogatives. One cannot expect an institution unilaterally to choose a path toward marginalization. As a result, it was better for the military that a civilian set the agenda on the border, as long as that agenda gave the country something beyond what Peru had offered previously in its insistence that the Rio Protocol provided all that was necessary for a resolution.

The second major point is that the time just after the 1995 war was ripe for the military to support a cooperative agenda. The "victory" of Tiwintza, after 150 years of defeats, was still fresh in the minds of the military and the populace. The consensus within Ecuador, however, was that Peru would not accept a low-level stalemate or defeat the next time. The Ecuadorean military was not anxious for a return to war, because the outcome was uncertain at best and the economic cost to the country would be totally unacceptable. Continuation of the border tensions thus put at risk the Ecuadorean military's hard-won prestige and its concern for national development.[19]

In addition, because the military had professionalized, military leaders were skeptical of the economic and strategic value of access to the Marañón (a tributary of the Amazon). The military's slogan "Ni un paso atrás" (Not One Step Back) could thus be maintained by emphasizing that an acceptable deal was not possible before Tiwintza and highlighting the contribution that the military had made to the country's development by facilitating a cooperative agreement. So while the civil-military relationship was still one of parallel spheres of influence, the military was ready to support an agreement if that resolution recognized the "victory" at Tiwintza and treated Ecuador with respect.

Two potential situations, however, could have led the military to interpret its interests in a way opposed to those who were pushing for an agreement with Peru. The military feared attempts by politicians to mobilize the military in support of specific political agendas and the rise of populism.[20] The Abdalá Bucaram government (1996–1997) raised the specter of both.

Frank Vargas Pazzos, retired air force general, ex-congressman, and advocate of using the military for extremely broadly defined civic action,[21] was minister of the interior in this administration. Vargas also led a military uprising against the government in 1986 which was forcefully put down by the army, thereby earning him its antipathy for tarnishing the military's reputation and for incurring a loss of life.

In addition, the military coup in 1972 was carried out partly in order to keep Abdalá Bucaram's uncle, Assad Bucaram, from winning the presidential elections scheduled for 1974. One of the points of negotiation for the transition to democracy in 1978–1979 was to keep the uncle from running. Abdalá Bucaram himself was seen as the successor to these populist movements and thus also a threat.[22] Consequently, if the military got drawn into domestic politics once again to "save the nation," it would have been difficult for military officers to appear less nationalistic on the border issue.

Peru

The civil-military dynamic in Peru was more complex than in Ecuador. It varied over time from military dominance in the 1960s and 1970s to civilian control in the 1980s. From the early 1990s, however, civilian dominance came at the cost of significant erosion in the institutional integrity of the Peruvian military establishment. The military's institutional capacity grew significantly in the 1950s and 1960s through concerted efforts to professionalize through training at the Centro de Altos Estudios Militares (CAEM, Center of Advanced Military Studies) and abroad, especially in the United States, and the progressive implementation of rigorous merit-based promotion criteria.[23] By the 1960s one of the manifestations of increased professionalism within Peru's military, especially the army, was the articulation of a national security doctrine that emphasized state-led national development and reform as the best way to avert social polarization and Communist-led political violence.[24]

When civilian governments faltered, as in an electoral impasse in 1962 and a mishandled nationalization in 1968, the armed forces assumed control of the government in nonviolent military coups. After the first takeover, international pressure led to the calling of elections a year later. In 1968, however, the military, perceiving that the civilian government was failing to implement promised change, came to power as an institution with a determination to remain in control until its reform agenda had been carried out.[25]

Over the course of its twelve years in power, the self-named Revolutionary Military Government (RMG) implemented major changes, including an agrarian reform, the nationalization of most important foreign investments, the establishment of worker-managed firms, a turn to the

Soviet Union for military training and major arms purchases, and a dramatic expansion in the state bureaucracy.[26] When faced with increasing differences within the military leadership over the course of the 1970s, the illness and removal of head of state Gen. Juan Velasco Alvarado (1968–1975), and a foreign debt crisis provoked by the inability to acquire the resources needed to continue financing the reforms, the regime decided in 1978 to return power to the civilian political establishment.[27]

Exhausted and disillusioned by its twelve years in power, the military oversaw open elections for a constitutional convention in 1979, which drew up the Constitution of 1979 that restored civilian rule through Peru's first elections with universal suffrage in 1980. Although the new constitution retained several of the reformist ideals of the RMG, the principle of civilian control over the military was also reestablished. Civilian control was maintained in Peru throughout the turbulent 1980s in spite of a progressive deterioration in the economy, including an inflationary spiral that impoverished millions and severely weakened the country's institutional fabric and a growing threat from the Shining Path guerrilla movement.[28]

In the face of such multiple crises, Peruvians turned away from the established political parties in the 1990 elections to choose Alberto Fujimori, an outsider with no prior political experience or party affiliation, as their next president. Even with an opposition majority in congress, Fujimori confounded expectations by embarking on an economic shock program to stem hyperinflation and introduce market reforms in order to restore confidence and economic growth. Without party allies, he depended on military support to back his government. He backed multiple changes that the armed forces were beginning to implement to confront an increasingly violent and aggressive Shining Path, but also moved to assert control over the military leadership. He did so by forcing the retirement of over twenty top army general officers known for their professionalism and selecting a head of the armed forces, General Nicolás Hermoza de Bari, on the basis of his personal loyalty rather than the merit criteria hitherto employed.[29]

The obvious cost of such reassertion of civilian control was a significant further erosion of the levels of professionalism of the Peruvian armed forces, even though this was masked for a time by the dramatic success against the Shining Path guerrillas. In September 1992 a small contingent of police tracked down and captured the group's leader, Abimael Guzmán Reynoso. This event, combined with other initiatives, marked the beginning of the end of the guerrilla threat to the country.[30]

The combination of the decline in the professional capacity of the Peruvian military, the failure to reequip the armed forces as a result of the prolonged economic crisis, and the focus on internal subversion made it virtually impossible to devote the attention or the resources needed to respond to Ecuador's advances along the northern frontier within the unmarked area of the border in the Cenepa River watershed.

Incentives of the Civilian Political Leadership

The central assumption of a rational choice institutional analysis is that political leaders respond to the incentives embodied in the political institutions within which they interact.[31] In democratic polities politicians face important constraints arising from the need to stand for periodic elections. The logic underlying this argument assumes that a politician's interest can be usefully condensed to winning elections. The claim is not that politicians have no other interests but that in order to accomplish whatever their goals are in politics they need first to be elected. The politician thus needs to offer the voters what they want in order to achieve electoral victory.[32]

To undertake a specific case analysis, we need a realistic if complex model of electoral politics in that nation. Chapter 3 concludes that Ecuador had an inchoate party system. *Ceteris paribus*, such a party system has negative implications for the country: the president cannot get support for policies in the legislature because the president's party neither controls congress nor has enough internal discipline to sustain policy coalitions.

These disadvantages for policy making were mitigated to a degree because the president was not constrained by the need to support the party in congress or prepare it for the next presidential elections. Such a president can try to make policy on its intrinsic merits or for very personal reasons, appealing directly to the public for support. Ecuador's constitution provided for a referendum when congress failed to pass legislation proposed by the executive. Going to the people presents its own challenges, however, as was demonstrated when President Sixto Durán Ballén lost the referendum for economic reforms in 1995.[33]

So how does this new incentive structure affect policy making in Ecuador? With reelection permitted beginning in 1996 and a weak party system, an Ecuadorean president would be more susceptible to breaking with tradition if medium-term benefits were likely; the payoff from a risky policy would need to occur by the time of renewed eligibility for office. The risk to the president (who is now betting on having the ability to win reelection) increases, so the payoffs need to be at least proportionately greater than they were before, when only considering the president's place in history. The medium-term benefits of viable resolutions of the Ecuador-Peru border dispute were not very great until 1995, largely because Ecuador had only a history of defeat at Peru's hands and because the internally oriented development strategies limited the possibilities for important economic cooperation.[34]

By the late 1990s, however, the Ecuadorean economy was sufficiently worse, regional economic liberalization had expanded dramatically, and nationalism had been enhanced by the "victory" at Tiwintza. As a result, the potential payoffs for resolving the dispute had greatly increased (in the forms of greater international investment, regional market expansion, and the opportunity to achieve international recognition as the political leader who negotiated the peace).

Bucaram was a political entrepreneur who probably could have succeeded in settling the border dispute had his term not ended prematurely. He began his presidency with high levels of popular support and, as suggested by this analysis, quickly used his administrative powers to implement controversial reforms.[35] These measures caused severe economic hardships for the working classes in the short term, but Bucaram's populism might have allowed him to survive until inflation decreased, the budget deficit was reduced, foreign investment increased, and economic growth accelerated. He was also laying the groundwork for a settlement with Peru by becoming the first Ecuadorean president to visit Peru and by suggesting to his counterpart there, Alberto Fujimori, that a mutual recognition of past errors would help the peace process.[36]

Bucaram might have followed the political trajectories of other surprise reformers in Latin America, including Carlos Andrés Pérez in Venezuela, Carlos Menem in Argentina, and Fujimori in Peru. Unfortunately, Bucaram's enormous political skills could not control his lust for power, the spotlight, and wealth. He filled his cabinet and state agencies with friends and family members. A new ministry was created to deal with indigenous affairs, directly challenging the independent political leadership of a key electoral ally.[37] Corruption, always a problem in Ecuadorean politics, reached new highs during his brief tenure. Bucaram's public behavior grew more outlandish with each passing day, and he seemed to relish the criticism that such actions provoked. When the end came, a man who had seven months earlier won one of the greatest electoral victories in Ecuador's history stood alone as millions marched in the streets, demanding his removal.

Bucaram's ouster demonstrated the willingness of Ecuadorean political elites and civil society to act on the margins of the formal institutional constraints when the stakes are high enough. On such a grave matter, the congress was willing to resort to extremes in order to use an obscure constitutional procedure (mental incapacitation) to depose the president. The Supreme Court, in turn, chose to stay out of the fight when the president contested congress's end run around an impeachment process that it could not win legitimately. The army initially supported constitutional procedures which favored Bucaram and subsequently his vice president but then resigned itself to treating the situation as a political rather than legal issue and stayed out of the controversy. Society simply rejoiced that Bucaram was gone, irrespective of how that had been accomplished.[38]

Congress placed its leader, Fabián Alarcón, in the presidency and called for new elections. The interim president attempted to pursue economic and political reforms but could get neither public nor congressional support. The economy began a significant deterioration under his watch (the GDP increased by only 0.4 percent in 1998 and would fall by 7.3 percent in 1999).[39] His sole accomplishment seems to have been to have won a referendum legitimating his replacement of Bucaram; even the Constituent Assembly and new constitution owed more to the ability of social movements to threaten the existing political order than to Alarcón's political vision.[40]

Jamil Mahuad won the 1998 elections and took office in August 1998 but did not have a mandate to undertake significant reforms even as both the political and economic systems seemed to be teetering on the verge of collapse. The failures of Bucaram and Alarcón indicated that an Ecuadorean president seeking to be a political entrepreneur on controversial issues should tread lightly.

Peru

Unlike Ecuador, where political turbulence was complicating and delaying progress on negotiations, in Peru President Fujimori presided over an increasingly consolidated political process between 1995 and 1998. Under the new 1993 Constitution, he won reelection decisively in 1995 and even squeaked out a narrow majority in the new unicameral legislature.[41]

Fujimori's electoral triumph at this juncture can be explained largely by his administration's success in overcoming the multiple crises that the country had faced when he came to office in 1990. Inflation had been reduced from over 7,000 percent to single digits, economic growth had been restored (in 1994 it was the highest in the developing world, at almost 12 percent), and Peru was once again in the good graces of the international financial community after restoring payments on the foreign debt.[42]

Of equal if not greater significance, however, was the Fujimori government's success in overcoming the threat of the Shining Path guerrilla movement. Through a series of measures, the government was able to restore peace to the country after more than a decade of increasingly generalized political violence.[43] Measures included a total readjustment in the military's counterinsurgency strategy, including support for local *rondas campesinas* (peasant militias or civil defense committees), the expansion of a specialized police unit to track and capture the Shining Path leadership, and a major microdevelopment program for the poorest districts in the country, overwhelmingly in the sierra highlands.

This combination of major policy adjustments, which together overcame the multiple domestic crises that Peru faced when Fujimori first took office in 1990, served to maintain his popular support in spite of the outbreak of a war with Ecuador that did not go well for Peruvian military forces. Economic liberalization benefited the coast disproportionately, microdevelopment achieved the same results for the sierra, and the defeat of Shining Path brought peace to the entire country. By cruising to an easy victory in the 1995 elections as a result of this dramatic domestic turnaround, Fujimori was in a position to negotiate from strength with Ecuador.

Immediately after his political victory, Fujimori moved to solidify his support with the military by declaring a blanket amnesty for any abuses committed in the course of the insurgency. Over the next three years he used his popularity, his dominance of the executive branch, and his majority in congress progressively to increase his control over the judiciary and the media. Nevertheless, what was to become his re-reelection project between 1998 and 2000, in violation of the constitution, did not figure prominently in the negotiation process with Ecuador.

While resolution of the border dispute would free Fujimori to concentrate on another electoral victory in 2000, his continued popular support (at least through 1998), Peru's stronger legal position with regard to the Rio Protocol, and a somewhat indifferent public opinion on the issue were more important considerations in dealing with the border issue. Fujimori's dominant leadership during most of the time when the negotiations with Ecuador were taking place enabled him to control the process on the Peruvian side and to impose himself on both the military and the foreign ministry leadership to secure an outcome that neither supported.

Conclusion

The argument in this chapter assumes that Ecuador had to make the major changes in its negotiating position for the Ecuador-Peru border dispute to be resolved. Peru needed to provide tangible benefits in return for such changes, but such Peruvian offers would be for naught in the absence of important changes in the Ecuadorean position. Our institutional analysis demonstrates the conditions under which the incentive structure of Ecuadorean politics could produce a significant change in the country's position on this issue.

The incentive structure confronting Ecuadorean presidents after 1995 was characterized by democratic institutions, ambivalent public opinion, a congress with a demonstrated ability to block policy, a weak party system, and a professional military that was also victorious in the conflict. The conditions under which this incentive structure could support a resolution were twofold. First, Ecuador needed a president who thought like a political entrepreneur and could see the benefits of resolving the border dispute for major programs of national social and economic development. As we have seen, given the institutional structure of Ecuadorean politics, a reformist president had to fight every step of the way to achieve them. Second, economic reforms needed to generate payoffs relatively quickly for the president to be able to overcome the multiple institutional resistances.

Could some other actor substitute for a president in Ecuador? If so, under what conditions might this occur? Our analysis suggests that civilian elites, diplomats, and the military were unlikely to lead in a cooperative direction on the border issue. In addition, at this time Ecuadorean civil society itself was too focused on internal social and economic concerns to perceive the dispute with Peru in a new light and demand changes.

As a result, the only possible alternative catalysts for a solution at this juncture would have been Ecuadorean nongovernmental organizations (NGOs) which focused on the border issue.[44] In the absence of dramatically increased costs to Ecuadorean citizens from the status quo (which would only occur in a major war), however, a reoriented public opinion was less likely to occur through NGOs, given the institutional constraints, than through an appropriately entrepreneurial president able to seize the moment.

This chapter's analysis of actor interests and strategic interaction suggests five conditions under which the likelihood of Ecuador's acquiescence to an agreement increased.

First, congress was unlikely to push for an agreement acceptable to Peru, but its approval of a settlement was essential. The constitutional change that permitted reelection meant that members of congress would be especially attentive to public opinion on this issue.

Second, public opinion in Ecuador had a nationalistic rather than an economic perspective on the dispute. In the wake of the "victory of Tiwintza" in 1995, however, public opinion could be persuaded to accept a package deal which subordinated territorial claims to various forms of development aid.

Third, while the military had historically opposed demarcating the border without attaining access to the Amazon/Marañón, its success in the 1995 war helped to overcome such opposition within its ranks, as did the Commerce and Navigation Treaty contemplated in the Rio Protocol, which guaranteed equal access and free passage to Ecuadorean shipping there as part of a settlement.

Fourth, the Ecuadorean situation was ripe for political entrepreneurship on the border issue; but, given the institutional constraints noted, such an entrepreneur would have to hold the presidency.

Fifth, this political entrepreneur would need outside help to increase an agreement's material benefits, thereby reducing the perceived costs.

Peru, in contrast, had a popular and dominant head of state in Fujimori, who had consolidated his power over the military by the time of the January 1995 war with Ecuador and soon thereafter also gained control over the legislature through the April 1995 elections. As a result, negotiated institutional adjustments in Peru's position were unnecessary: the president was in a position to call the shots, once the dust had settled on the conflict itself. The key for Peru came at the very outset of the war, when Ecuador accepted a return to the framework of the Rio Protocol after thirty-five years of denying its validity.

This change in the historic Ecuadorean position opened the way for negotiations within the parameters of the protocol, with the assistance of the guarantor countries. As a result, it was not necessary for Peru to do more than recognize that a problem existed and that negotiations with Ecuador were necessary to resolve it. The drawn-out nature of the negotiation process, due to a combination of multiple circumstances and internal institutional necessities, served to enable Ecuador to achieve the essential adjustments required to resolve the issue on terms quite at odds with its historic position.

6 HEMISPHERIC DIPLOMACY AND THE POLITICS OF A SOLUTION

Diplomacy is an ancient tool used by states to promote cooperation as well as war. Perhaps diplomacy's greatest challenge lies in overcoming the residues of distrust produced by the existence of anarchy in the international system. States may want to cooperate, but they need to consummate bargains on important issues to bring that cooperation to fruition. The priority that states give to those issues makes it fundamental that the credibility of goodwill and the transparency of cooperation win out over efforts to maximize gains by deception and treachery.

We have seen thus far that prior diplomatic efforts failed to resolve the Ecuador-Peru conflict, even when high-prestige individuals became involved. Over the years at least thirteen major initiatives to solve the problem failed to do so. Some involved periodic arbitration efforts by the pope, the Spanish Crown, and the president of the United States. Others included a number of draft treaties between the parties that, with the exception of the 1942 Rio Protocol, were never ratified.[1]

The controversy never seemed to be quite ripe for resolution. During most of the nineteenth century the dispute was not sorted out between the two governments because they considered a vast, remote, and virtually unpopulated jungle to be unimportant. In the twentieth century, with an early rubber boom and the later discovery of oil in the region, along with growing nationalism and a sense of national identity, no resolution was reached because the parties saw it as too important.

With the 1995 war, however, the internal, regional, and international contexts became more conducive to a diplomatic approach to the dispute. Ecuador's success on the battlefield gave both the military and the general population a new sense of national pride that tempered the humiliation of past defeats and opened the government to negotiations within the provisions of a 1942 treaty with Peru that it had eschewed for decades. For its part, the Peruvian government, though unexpectedly stymied militarily, saw an opportunity for diplomacy to achieve the definitive boundary settlement that it believed the 1942 treaty offered. The specter of a full-scale war with modern weapons if the dispute persisted also contributed to both parties' willingness to consider a negotiated alternative. The prompt action of other Western Hemisphere governments in calling for an end to hostilities and offering their assistance in finding a diplomatic solution reinforced the more favorable context that was emerging.

This chapter details how the institutional structure of the negotiations and the diplomatic skill of key individuals enabled Peru and Ecuador to take advantage of this historical opportunity, in spite of the deeply etched historical legacy of bilateral conflict over Latin America's oldest border dispute, unexpected domestic problems, and very different negotiating approaches. It also explains how these key factors produced a definitive resolution instead of sliding back into confrontations or even a new war.

The International Legal Context

The boundary dispute between Ecuador and Peru involved the longest-standing multilateral mechanism for international conflict resolution in the region. It was established (as detailed in Chapter 2) in the 1942 treaty generally known as the Rio Protocol to bring an end to a full-scale war in 1941, which Peru won decisively.[2] The treaty set out the terms by which four "guarantors"—Brazil, Argentina, Chile, and the United States—would assist the parties to mark the boundary and help them work out any disagreement in the process. It brought to bear an international presence on the process, bringing a higher level of concern to conflict resolution for a problem that tended to be seen as of only minor importance outside the immediate region. It also provided a vehicle for bringing suggestions and alternative approaches to the table for consideration by two parties with a deep-seated distrust of each other.

Although the treaty was heralded at the time as a final settlement of the border problem, the guarantors expected that their involvement after 1942 would be short-term, a few years at most, and largely technical in nature. Nevertheless, the guarantors were still at work over fifty-three years later—indeed, facing their most daunting challenge. For under the Rio Protocol, their role ended only when the last boundary marker was set in place. Given the deeply embedded differences between the parties to the dispute, the involvement of the guarantors proved to be absolutely essential to its ultimate resolution.[3] The key points of the document note the continuing collaborative and assisting role of the guarantors but recognize that ultimate responsibility for resolution rested with the parties. The 1942 treaty, in sum, contained the necessary provisions under international law for a diplomatic solution—as long as Ecuador and Peru could agree to it.[4]

Although implementation proceeded largely as envisioned for several years and all but a small segment of the border was marked, geographical anomalies found after aerial mapping plus a more nationalistic political climate in Ecuador led to suspension of participation in demarcation between 1948 and 1950 and to a unilateral declaration in 1960 that the protocol was "null." As long as Ecuador took this position, which it did for more than three decades, no further progress could be made under the terms of the protocol. Furthermore, many governments after 1948 used the slogan "The Amazon is ours" to whip up popular support and reinforce nationalistic sentiment, making it very difficult to approach the issue with any degree of objectivity.[5]

Given such long-standing opposition to the validity of the Rio Protocol by various Ecuadorean governments, a January 1995 announcement by President Sixto Durán Ballén (1992–1996) took everyone by surprise. Just as a new armed conflict with Peru was escalating, he stated that Ecuador would welcome the assistance of the guarantors within the purview of the protocol. Whatever the motivation, this major shift in Ecuador's position reestablished the Rio Protocol as the international legal instrument within which discussions toward finding a solution to the dispute could begin.

The Application of the Rio Protocol

The central challenge for the guarantors at this historic juncture, then, was to provide the assistance contemplated within the Rio Protocol to help the parties overcome their long-standing and deeply held differences on the border issue and to find a specific set of solutions within the protocol's provisions. Over the years both Peru and Ecuador had developed positions on the dispute that had become increasingly intractable. They came to the negotiating table in 1995 with what amounted to non-negotiable stances. For Ecuador, no solution would be acceptable that did not include sovereign access to the Amazon River. For Peru, no solution would be acceptable that did not adhere to international law and define the boundary in terms of the precise points specifically laid out in the 1942 Rio Protocol. Such positions on both sides had wavered little between 1948 and 1995, whether under military or civilian governments, and had thwarted repeated attempts by representatives of the parties and the guarantors to find some mutually acceptable resolution of the issue.

Given its exceedingly long history, the multiple failures of both bilateral and multilateral efforts to resolve it, and intense nationalistic sentiments expressed on both sides of the border, how can the successful resolution of the dispute be explained? The discussion that follows lays out the various elements that contributed to the peaceful settlement that eventually emerged.

The only realistic option was to conduct negotiations within the parameters of the Rio Protocol, but this was not possible as long as one of its signatories denied its validity. The results of the short border war in 1995, however, gave Ecuadorean decision makers the ability to shift position after thirty-five years, permitting the Rio Protocol once again to serve as the framework for resolving the long-festering border dispute. Without Ecuador's historic decision, no forward movement could have occurred, whether or not it ultimately proved successful.

By reapplying the Rio Protocol, the guarantor mechanism, one of its crucial components, could once again come into play. Although under the treaty the guarantor countries were authorized only to assist and not to arbitrate, they could in principle play a decisive role in moving the parties forward. In the process, they could also serve as a kind of lightning rod to deflect public criticism from positions taken in the negotiating process by the parties themselves in ways that enabled the process to stay on track.

Given the deep-seated distrust between the parties, however, the guarantors had little realistic hope of success without strong individual abilities, a clear sense of how to proceed, and a willingness to endure the tedium of their day-to-day work. As it turned out, the guarantor representatives were a very compatible and cohesive group. They were able to retain their legitimacy and even enhance it over time by consistent adherence throughout the entire process to five stated principles that they adopted at the outset as their official guidelines (see Chapter 2). Within these principles, the process of working toward a diplomatic resolution of the dispute passed progressively, if fitfully at times, through three distinct phases. Each had to be achieved before it would be possible to move on to the next.

Stage 1: Establishing a Multilateral Peacekeeping Mission

In spite of the efforts of the guarantors, hostilities continued through February, ending only after Peru and Ecuador agreed on February 28 in

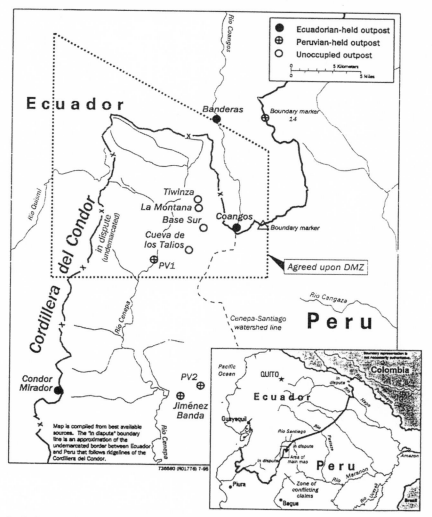

Ecuador/Peru: Demilitarized Zone (1995). Courtesy of the Office of the Geographer, U.S. Central Intelligence Agency (re-drawn by the North-South Center Press).

Montevideo, Uruguay, to commit their countries to peace. At that point, the provision in the Itamaraty Peace Declaration to create the Ecuador-Peru Military Observer Mission (MOMEP) could go forward.[6]

MOMEP included small contingents from each guarantor country as well as from Ecuador and Peru and was financed by the parties to the conflict themselves. From the outset, a Brazilian general officer served as "coordinator" of MOMEP, with a U.S. colonel as the head of the American contingent. During the first year the United States provided the complex logistical support as well. As the peace process dragged on, however, logistics passed to the Brazilians in their role as overall coordinators under the protocol.

This multilateral military presence played a vital role during the initial months before diplomatic efforts got fully under way. Between March and August 1995 MOMEP was able to achieve the separation of contending forces and their withdrawal from the disputed area as well as the establishment of a demilitarized zone where the fighting had taken place.

This MOMEP contingent also played a major role much later in the peace process, at a critical moment in late July and early August 1998, just as the elements for resolution were being worked into their final form. Peruvian forces, frustrated by their losses in 1995 and by the lack of definitive diplomatic resolution after more than three years, were preparing for a preemptive armed assault at the border, alleging provocations by Ecuadorean forces. At that juncture, the presence of the multilateral force helped to defuse a situation that could easily have caused the entire peaceful resolution process to unravel.

Stage 2: Defining Issues and Building Mutual Confidence

With the military situation stabilized by late 1995 thanks to MOMEP, by early 1996 the guarantors could begin ministerial-level discussions in Lima and Quito to assist the parties to identify the outstanding points of disagreement on each side and to set up the procedures to follow in future substantive negotiations. Remarkably, over all the years of tension and periodic confrontations, neither Peru nor Ecuador had ever specified in precise terms exactly what the points of disagreement were between them.

At this point, the guarantors hoped that enough momentum could be generated to work out a solution before Ecuador's national elections on May 7, 1996. On January 17 and 18 Ecuador's foreign minister, Galo Leoro, met in Lima with his Peruvian counterpart, Francisco Tudela, along with representatives of the guarantors. This meeting, as anticipated, focused on procedural matters, with "accords . . . on continuing the peace process, the site of the discussions, the structure of the delegations, the confidential nature of the talks, the role of the guarantor nations, and the need to extend the operation of MOMEP."[7]

On February 22 and 23 the parties met again, in Quito, to continue discussions of procedural matters. This meeting occurred after other gatherings had taken place—a meeting of military representatives of Peru and Ecuador in the disputed area on February 10 and 11 concerning a possible nonaggression pact and avoidance of an arms race and a meeting of the parties and the guarantors in Brazil on the occasion of the first anniversary of the cease-fire. The parties agreed to create a bilateral commission to oversee arms purchases and a joint military working group to promote security and stability in support of the diplomatic negotiations as well as to adopt measures to lower the risk of troop clashes along the border.

As an outcome of these first meetings, the parties took the significant and historic step of setting down in writing for the first time their remaining substantive impasses concerning the boundary and exchanging their lists (see map, page 109).[8] In spite of mounting tensions as the process went forward, with claims and counterclaims of over-flights, troop movements, and demilitarized zone incursions, by March 1996 each party had recorded its respective concerns.

For Ecuador, these included the following:

1. Partial inexecutability of the Rio de Janeiro Protocol due to the absence of a watershed between the Zamora and Santiago Rivers. Free access and Ecuadorean sovereignty to the Marañón-Amazon.
2. Boundary demarcation problems:
    a. The Cuzumaza-Bambuiza/Yaupi sector. [Ecuador claimed that this ridge was not part of the Cordillera del Cóndor because of differing rock composition.][9]
    b. The Lagartococha-Güepí sector. [Ecuador claimed a problem here due to an issue of international law stemming from the Braz Dias de Aguiar arbitration.]
3. Problems produced by the intersecting of the rivers by the survey lines. [Identified as three areas in the Pastaza, Tigre, and Curaray river sectors.] Problem on the Napo River in the Yasuní-Aguarico sector.
4. The Zarumilla Canal.[10] [A blockage issue—too much silt blocking the canal's water flow.]

Peru described its substantive differences in the following manner:

For Peru, as Ecuador knows, the phrase "enduring resolution of the substantive differences" means completing the demarcation of the boundary line as established in Article 8 of the Protocol of Peace, Friendship, and Boundaries subscribed to in Rio de Janeiro on January 29, 1942, in conformity with its complementary provisions and with the Award of the Brazilian Arbiter Captain Braz Dias de Aguiar.

For Peru there are two sections of the border where the demarcation differences may be found:

1. In the Lagartococha sector:
   a. The source of the Lagartococha River-Güepí River.
2. In the Cordillera del Cóndor:
   a. Between the boundary marker "Cunhuime Sur," point "D" noted in the Dias de Aguiar Brief (on the Cordillera del Cóndor from the point along the Zamora-Santiago height of land where the spur juts out) and the boundary marker "November 20."
   b. Between the boundary marker Cusumasa-Bumbuisa and the confluence of the Yaupi and Santiago Rivers.[11] [Peru claims that this ridge, running adjacent to the Santiago River, is part of the Cordillera del Cóndor.]

The most intractable point of Ecuador's list was that the only acceptable solution must include "free and sovereign" access to the Amazon. Peru's position was equally firm in its insistence that the boundary be defined along the watershed of the Cordillera del Cóndor as set out in the Rio Protocol. Nevertheless, the formal articulation of the remaining impasses served to specify the areas of concern that had to be dealt with when the parties entered substantive negotiations.

Finding some acceptable set of procedures within which Peru and Ecuador could begin to negotiate was the next major challenge that the guarantors had to face. This sensitive topic was first discussed in Buenos Aires in June 1996 and finally agreed to by the conclusion of a follow-up meeting in Santiago in October. The meeting was exceedingly contentious and difficult, according to participants;[12] but after the dust settled, the parties did come to an agreement on the following key points:

1. Substantive discussions to begin before the end of 1997
2. Discussions to be "continuous" until final resolution
3. Partial decisions to be final only after agreement on all points
4. The parties to specify points of agreement and disagreement

5. Guarantors to "enforce" agreements and to propose solutions where the parties can't agree among themselves.[13]

This accord, however difficult it was to achieve, permitted the conflict resolution process at last to move on to the third stage, substantive negotiations between the contending parties.

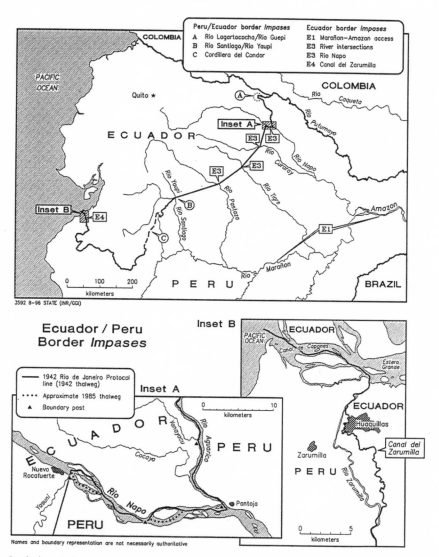

Ecuador/Peru: Border Impasses (1996). Courtesy of the Office of the Geographer, U.S. Department of State (redrawn by the North-South Center Press).

Stage 3: Proceeding with Substantive Negotiations

The first meeting to enter into negotiations under the rubric of the Santiago Accord was scheduled for Brasilia in early 1997. Unforeseen events
in both Peru and Ecuador, however, temporarily scuttled the diplomatic
schedule.

In Peru, the Tupac Amaru Revolutionary Movement (MRTA) took over
the Lima residence of the Japanese ambassador to Peru on December 17,
1996, just before substantive discussions were to begin in Brasilia (December 20). Among the hostages was Peru's foreign minister, Francisco
Tudela, whose participation in the negotiations with Ecuador to that
point had been constructive by all accounts and whose continuing role
was perceived as vital for any ultimately successful resolution of the
dispute.

In Ecuador, elected President Abdalá Bucaram was voted out of office on February 14, 1997, by his country's congress for "mental incapac-
ity" after only six months on the job. In the course of his truncated term,
whatever his foibles and mistakes at home, he was committed to solving
the border problem once and for all. To this end, he made an impromptu
if symbolic visit to the border only days after taking office and mingled
with Peruvians. In addition, he quickly established good rapport with
his Peruvian counterpart, Alberto Fujimori, and made the first state visit
to Peru ever by an Ecuadorean president, just a month before his ouster.

These events delayed negotiations for several months but gave time
for the guarantors to work out a procedure for separating the key issues
into discrete components. They concluded that an overall solution was
more likely if the major concerns could be dealt with in four separate
negotiations through meetings in the guarantor country capitals, headed
by the country's representative and with the parties forming separate negotiating teams for each venue. When the postponed discussions finally
resumed in Brasilia in April 1997, Ecuador and Peru agreed to this arrangement; in September and October the parties were finally able to begin formal negotiations within the Santiago Accord's parameters in each
of the guarantor capitals.

In Washington, D.C., meetings covered border integration and external financing; in Buenos Aires, the Treaty of Commerce and Navigation;
in Santiago, confidence-building measures; and in Brasilia, differences
related to the border demarcation itself. The third and critical phase was
finally under way.[14] By separating the issues into their discrete parts, the

parties made substantial progress over the next few months and were able to reach agreements in three of the four capitals. Only the border demarcation problem remained to be resolved.

An important breakthrough was achieved in Buenos Aires. There Peru and Ecuador agreed on procedures to follow in the critical border demarcation negotiations in Brasilia, with Peru supporting a definitive arbitration by the guarantors while Ecuador proposed arbitration by an outside "eminent personage" in case of differences between the parties.[15] The willingness of both parties to consider arbitration in some form at this point in the negotiations suggested their mutual recognition that the "non-negotiable" portion of their stated impasses might have to be subject to resolution by outside parties. This adjustment in the two countries' positions, however belated it was, turned out to be a crucial component in the final stages of the dispute resolution process.

Nevertheless, two major stumbling blocks still had to be overcome. One was procedural. Could a mechanism be found (such as some form of outside party or guarantor review) to ensure resolution even if the parties could not agree? The other was substantive. Both the Peruvian and the Ecuadorean lists of remaining differences contained one point that did not appear to be negotiable.

The Peruvian list included the demand "to complete the demarcation of the border as established . . . [by] the Rio Protocol." Ecuador's list included the assertion that the Rio Protocol was "partially inapplicable" and that Ecuador must have "free and sovereign" access to the Marañón-Amazon. Peru was believed to have communicated to the guarantor countries its unwillingness to negotiate these points.[16] In effect, at least for public consumption, both parties adopted a non-negotiable position in one critical respect which appeared to leave the guarantors with very little maneuvering room to assist the parties to find common ground for final resolution of the dispute.

Nevertheless, a way through this crucial remaining impasse was found in the mechanism of the expert commissions, set up by the guarantors with the agreement of the parties and charged with preparing reports on the impasses for submission to each delegation involved in the negotiating process. The brief provided by the map experts on the boundary commission on May 8, 1998, proved to be the most important. Although its findings remained secret, information leaked out that, although several minor border issues favored Ecuador, they confirmed the long-standing Peruvian position that the watershed stipulated in the

1942 protocol indeed existed along which the boundary should be drawn in the disputed area of the Alto Cenepa: the Cordillera del Cóndor.[17]

Given this conclusion by the experts, the crucial question was now (as it had been all along) just how to get both parties to agree to this line. In the following delicate and difficult negotiations, Peru acquiesced to a slight border adjustment in Ecuador's favor in another small area in dispute, in the Cuzumaza-Bambuiza-Yaupi zone. But Peruvian negotiators were not willing to concede any territorial adjustment along the Cordillera del Cóndor. Their Ecuadorean counterparts, believing that their country had already made a major concession in giving up insistence on sovereign access to the Amazon, were not willing to accept the Peruvian position that the watershed along the Cordillera del Cóndor was the only acceptable boundary line in the undemarcated area of the Alto Cenepa. This would mean, from an Ecuadorean perspective, that they would have nothing to show for their military successes against Peru in the 1995 conflict, which would be unacceptable for key actors at home.

So the historic impasse remained. The reintroduction of some troops into this area by both Peru and Ecuador created a new crisis during the first days of August 1998, which could easily have provoked an all-out war and the total collapse of the negotiation. This may well have been what some of the more nationalistic military and civilian elements on both sides were seeking. At the height of the crisis, however, President Fujimori insisted that there be no military reaction by Peruvian forces. MOMEP also assisted in defusing the situation by agreeing temporarily to expand its purview and to oversee troop withdrawal. This permitted a return to the negotiating table rather than succumbing once again to the use of armed force.

The impasse was broken by a deft diplomatic stroke by which, after multiple president-to-president meetings, the parties agreed to disagree on an Alto Cenepa solution (while accepting the negotiated agreements on the Treaty of Commerce and Navigation, confidence-building measures, and border integration) and leave definitive resolution in the hands of the guarantors.

Furthermore, the parties agreed in advance (at guarantor insistence) that they would accept the guarantors' determination on the Alto Cenepa as final by submitting the arbitration proposal to their respective congresses. Both congresses met in their respective capitals at the same time and accepted the measure after extended discussion on October 23. Now it became possible for binding arbitration by the guarantors to become the instrument for a definitive agreement.[18]

Peru and Ecuador had agreed in advance to accept the following key provisions of the guarantor decision:

1. To draw the boundary along the Cordillera del Cóndor as indicated in the map experts' brief
2. To establish a demilitarized nature preserve on both sides of the border thus established, under the administrative control of specially trained personnel from each country within its respective territory
3. To give Ecuador's government control in perpetuity as a private owner of a small parcel of land (one square kilometer or 250 acres) at Tiwintza, within Peruvian territory, where Ecuadorean authorities could erect a monument to honor their fallen forces in the 1995 conflict and to which they would have unimpeded access from their side of the border over Peruvian territory.[19]

The burden for the resolution of the non-negotiable negotiating positions of the parties thus fell to the guarantors, whose solution brought a definitive resolution to a dispute that had long eluded the best efforts of their predecessors. On October 26 the presidents of Ecuador and Peru signed the historic agreement, in November the Ecuadorean and Peruvian congresses ratified the agreement and the Treaty of Commerce and Navigation, and in mid-December the instruments of ratification were exchanged, finalizing the historic agreement. Peace was at hand, at last!

In assessing this long and difficult process over three and a half years of "getting to yes," the most important elements contributing to overcoming the weight of history surrounding the boundary dispute include the role of the guarantors, the particular skills of its members at critical moments, their ability to maintain a unified approach, and the role of the parties themselves. Absent any one of them, it is hard to envision how success could have been achieved. In combination, however, they produced the success that had been so elusive for so long.

The Role of the Guarantors

It is quite unlikely that the border dispute, given its long historical trajectory and the entrenched positions of Ecuador and Peru, would have been resolved without the participation of the guarantors. They took their responsibilities seriously, were committed to doing everything they could to help the parties solve the problem once and for all, and were prepared to devote the energies and resources necessary to that end. The successful establishment and continued operation of the multilateral military operation (MOMEP) in and around the disputed border area was a case in point. Another example of guarantor countries' commitment was the willingness of the U.S. State Department to second to the mission a senior person of ambassadorial rank and a more junior assistant virtually full-time as well as to provide short-term support from other appropriate agencies and offices when needed. In addition, the other guarantors named experienced vice ministers of foreign relations of ambassadorial rank. Brazil assumed its usual coordinating role and hosted some twenty guarantor country meetings in Brasilia, many with Ecuador's and Peru's participation as well. Argentina and Chile were also actively involved, including the hosting of the key confidence-building meetings in Buenos Aires and Santiago in 1996 and the discrete issue negotiations in 1997 and 1998. Furthermore, all guarantor representatives were well regarded by their peers and were taken seriously by the parties themselves as well.

The ultimate success of the guarantor mission's labors was enhanced by multitrack diplomacy initiatives that included support for conferences of leading academics and civic leaders from both countries and mixed civilian-military groups, which combined to help weave networks favoring moderation and peaceful resolution. Military-to-military contacts through MOMEP and new U.S. bilateral military-to-military initiatives reinforced such networks in key institutions and among their leaders. Together such initiatives progressively improved the bases for constructive engagement between the parties and made even more self-evident the negative consequences of maintaining a boundary dispute without resolution.[20] The contact, consultation, and transparency which resulted helped to reinforce the diplomatic efforts and to provide a groundswell of support among influential constituencies over time that contributed to the legitimization of the negotiation process and to its successful outcome.

The Special Skills of the U.S. Guarantor Representative

While all of the guarantor representatives were highly qualified individuals, the particular abilities of one, ambassador Luigi Einaudi of the United States, stood out. Working quietly behind the scenes, he gradually earned the trust and respect of the parties.[21] Shortly after the conflict broke out, he was approached by assistant secretary for inter-American affairs Alexander Watson to take on the guarantor assignment and quickly accepted. He spent the next four years working almost full time on the issue, two of them at the explicit request of both parties after he had officially retired from the State Department. He was the only guarantor representative who could spend this kind of time on the task. The others were distinguished and respected professional diplomats as well, but all had continuing parallel responsibilities in their own foreign ministries that limited the attention they could give to the border problem. Partly because of this limitation, in fact, they designated Ambassador Einaudi as the guarantor intermediary representing the body between their formal meetings. This in effect made him the first among official equals.

By all accounts, Einaudi played an indispensable role throughout the process. He is credited by various key actors with everything from finding ways to express issues that would not offend either party (such as the "substantive impasses" designation), to coming up with the private property idea for Tiwintza that opened the way to the breakthrough needed for resolution.[22] Both parties came to trust his judgment, his fairness, his scrupulous adherence to appropriate procedures, and his discretion. The general conclusion among participants is that without his persistent, sometimes tenacious, involvement the Ecuadoreans and the Peruvians would never have signed and ratified an agreement.[23] Ambassador Einaudi's participation confirms the adage that diplomacy transcends cold calculations of costs and benefits, balance points, and game theory—individuals do make a difference.

A Unified Approach

Another important component of the guarantor dynamic as the dispute resolution process moved forward was the members' positively reinforcing rapport. This often enabled them to move forward as a single body in their interactions with the parties, even though the United States appeared to be the guarantor most interested in a rapid resolution and Brazil the most willing to let the diplomatic process work at its own pace. On this occasion, guarantor representatives brought less of their respective countries' foreign policy agendas to the table than in the past.

Several pre-1942 initiatives, in fact, had to be abandoned when the parties and their citizens mounted major protests against what they perceived as attempts to impose outcomes more aligned with the outside participants' interests than their own.[24] During the 1950s as well, when the Rio Protocol was in effect and the guarantors had been regularly called upon to resolve periodic border incidents, the regional foreign policy dynamics of the time had worked against a unified guarantor position on the issue. Argentina tended to favor Peru, while Chile frequently acted on behalf of Ecuador. In addition, rivalry for regional influence between Brazil and Argentina at this time often produced tension among them in guarantor meetings and impeded the dispute resolution process.

A further difficulty during the 1950s, when the parties regularly requested guarantor assistance to help resolve incidents, was the slow and passive nature of the response. Guarantors proved reluctant to make decisions on their own, delayed resolutions until foreign ministries had vetted drafts, and bent over backward to avoid antagonizing either party. As a result, final statements tended to be so watered down that little progress could be made, with the delays often contributing to the escalation of minor incidents into full-blown crises.[25] Such problems were generally avoided once the guarantor mechanism was reinstated in 1995, although all was not smooth sailing on this occasion either. There were revelations of Argentine sales of military equipment to Ecuador through Venezuela and Panama during the fighting and concern that the traditional enmity between Chile and Peru (dating from Peru's defeat in the War of the Pacific in 1879–1883) might affect Chile's views as a guarantor.[26] Nevertheless, these issues failed to provoke the kind of reactions that had scuttled peace initiatives in the past, largely because the guarantors on this occasion were able to hew to their unified approach to the process.

The Role of the Parties

In this round of negotiations, both Ecuador and Peru had their own reasons to be amenable to outside assistance. Peruvian diplomats had concluded that the guarantors would favor their international law–based interpretation of the most appropriate solution. Ecuador's counterparts felt that there was a good possibility of a third-party judgment in favor of their historically based claim.[27] At the same time, however, long-standing positions were retained. Ecuador could not accept a final agreement that did not include a territorial concession by Peru in the long disputed area that had not yet been demarcated. Peru could not accept any resolution that did not place the boundary at the height of land of the Cordillera del Cóndor watershed, as stipulated in the Rio Protocol.

Working through this major impasse was a complex and multifaceted process that involved adjustments on both sides and faced various obstacles along the way. Peruvian diplomacy was thrown off stride at a critical point in the negotiations, with the surprise resignation of foreign minister Francisco Tudela in mid-July 1998 due to his opposition to President Fujimori's domestic political maneuvers, thereby removing Peru's most credible spokesperson from the border dispute negotiating process. His replacement in August by Dr. Eduardo Ferrero Costa, a knowledgeable international relations academic from a prominent political family but without government experience himself, set the process back even more. Ferrero Costa came to be seen over the succeeding months as a less flexible negotiator than his predecessor and as one who did not have Tudela's independent stature.[28]

Change in the leadership of Peru's Negotiating Commission, however, had a different effect. The replacement of Ambassador Alfonso Arias-Schreiber, a distinguished retired military officer and diplomat known for his hard line on the border issue, with Dr. Fernando Trazegnies, a well-known professor of international relations at the Catholic University, introduced a more moderate perspective to that important body from the Peruvian side, even though he, too, lacked an independent base of support. The overall result was a process over which President Fujimori could exercise greater personal control, which turned out to be a positive development: it became increasingly evident that he had become committed to a full and final resolution of the dispute.

Ecuador's interim president Fabian Alarcón, who was expected to be little more than a transitional caretaker calming domestic waters after the tumultuous regime of President Bucaram, strengthened both his own internal position as well as his country's in the discussions with Peru by winning a national referendum in May 1997 confirming his position as interim head of state until elections scheduled for May 1998. Foreign minister José Ayala Lasso's experience and apparent ability to gain independent control over the intra-Ecuadorean position in the bilateral discussions also gave Ecuador an advantage. The country was further aided by the presence of ambassador Edgar Terán as head of its Negotiating Commission. He was Ecuador's former ambassador in Washington and a person closely involved in discussions on the issue from the beginning of the 1995 conflict.

Such coherence and professionalism within Ecuador's negotiating team undoubtedly contributed to a more pragmatic stance regarding the country's historic position and produced a critical shift in January 1998, when Minister Ayala Lasso dropped the word "sovereign" for the first time in a public speech. By declaring that a goal of the negotiations was "access to the Amazon," the phrase used in all subsequent official statements, Ecuador was signaling its readiness to find some solution to the impasse short of the nation's historic aspirations.

In a follow-up television interview several months later, the foreign ministry general secretary, Diego Ribadaneira, articulated what that solution might entail and how it could still accomplish the basic objective. He explained Ecuador's position as one that retained the goal of full access to the Amazon region but that would be accomplished by the Treaty of Commerce and Navigation within the parameters of the Rio Protocol.[29] To the guarantor community, these official Ecuadorean foreign ministry statements signaled that Ecuador had found a way to reconcile historic aspirations within international legal mechanisms and that negotiating space within them was now available for working toward a final settlement of the dispute. Ecuador's more flexible position shifted the burden for a reciprocal adjustment of historically non-negotiable positions to Peru.

The July 12, 1998, election victory of Jamil Mahuad as president of Ecuador over Jaime Nebot, in a closely contested runoff, turned out to be another major development. Even though the campaign, by mutual agreement, had scarcely mentioned the border dispute, it was widely believed that the victor held a more moderate position on the issue than his rival. Between the election and the inauguration, however, new Peruvian military incursions into part of the disputed area in late July and early August threatened to provoke a renewal of hostilities. The president-elect nonetheless chose to make the border question his first order of business and offered to meet with his Peruvian counterpart to seek a negotiated peace in his inaugural address on August 10. The Peruvian foreign minister's immediate negative response to the new Ecuadorean president's olive branch only heightened the perception that a new outbreak of hostilities was imminent.

At this critical moment, a combination of developments averted disaster and brought the conflict back to the negotiating table. The Argentine foreign minister, Guido di Tella, visited Lima on his way back to Buenos Aires from President Mahuad's inauguration. Exercising his role as a guarantor representative, he succeeded in securing an agreement from President Fujimori that Peru would indeed return to negotiations. The Peruvian president immediately imposed his authority on his own hard-line foreign and defense ministers, and MOMEP was then able to extend its purview temporarily beyond the demilitarized zone to secure the withdrawal of Peruvian and Ecuadorean troops from the area. And in addition to Argentina's vital role in Lima, the good offices of the other guarantors, especially Brazil as coordinator, included President Fernando Henrique Cardoso's extensive personal diplomacy with both the Peruvian and the Ecuadorean heads of state at critical junctures.[30] A parallel track of private diplomacy was also starting at this time. One of President Mahuad's closest advisors persuaded him to invite his Peruvian counterpart to lunch at a neutral site. Their four-hour private repast in Brasilia in early August 1998, with aides sitting outside and unknown to the public at the time, proved foundational to the peace that would later emerge.

With President Fujimori now committed to a resolution of the dispute, it became a question of finding the right formula which would meet Peru's demand for a solution within the terms of the Rio Protocol. In August he forced the resignation of the long-term commander in chief of the armed forces, Nicolás Hermoza de Bari, believed to oppose the terms of a final settlement with Ecuador as they were emerging; Hermoza de Bari was also thought to have allowed the July 1998 infiltration of Peruvian military forces back into the disputed area of the border. As the negotiating situation evolved in August and September 1998, Fujimori increasingly took charge of the day-to-day strategy. This provoked the resignation of Minister Ferrero Costa early in October, on the eve of the final agreement, both because he felt that he was being shunted aside in the negotiations and because he had always opposed a territorial concession of any kind to Ecuador.

Throughout August and September presidents Mahuad and Fujimori had followed up on their initial private lunch with a series of meetings. In spite of the positive dynamic that the presidents' meetings created, the two heads of state could not overcome one last hurdle. President Mahuad felt that he could not take to the people of Ecuador a boundary document that did not include access and territory on the Peruvian side. President Fujimori felt that he could not take to the people of Peru one that did.

So the two presidents agreed to disagree but at the same time brought their political skills and newfound rapport to bear to find a way out of this final impasse. They turned to the guarantors and agreed in advance to accept their solution in what amounted to binding arbitration. Their personal diplomacy, along with the presence of a preexisting mechanism to which they could turn (the guarantors), paved the way for the historic agreement that provided a definitive resolution of the dispute.

The Role of Internal Political Dynamics

Finding a comprehensive solution to the Ecuador-Peru border dispute confirms the maxim that the last mile of a marathon is the most difficult for all participants. In this particular case, many of the challenges faced in the final stretch arose from internal political dynamics in both countries. One challenge was posed by internal events quite apart from bilateral issues that delayed the peacekeeping negotiating process for several months (as described above). Another related to the distinctive features of domestic politics in each country.

Peru's internal political dynamics were much more stable and predictable than those in Ecuador. President Fujimori, in power since 1990 (which he consolidated with a self-initiated coup in 1992, a new constitution in 1993, and his reelection in 1995 by a significant margin), was firmly in control of the government, had the military behind him, and had a pliable majority in congress. This was confirmed by an unrelated development, congress's August 1998 rejection of a referendum on whether he could run for a third term in office, even with more than the constitutionally mandated number of signatures required for its submission to a national vote. Ecuador's democracy, in contrast, was weak and fragile, with a legislature that was almost always controlled by different parties than occupied the executive branch. Over the course of the 45-month effort to defuse and solve the border problem, Ecuador had four presidents, symbolizing the instability of domestic politics there.

Such differences in internal political dynamics led to radically different approaches in the way each government dealt with domestic constituencies on the status of the border dispute negotiating process as it proceeded. Ecuadorean officials consulted regularly with key elites, including congress, the military, business, and the media, at each step of their discussions with Peru. Even though this slowed down the international negotiation process, to the consternation and frustration of all involved, important sectors of Ecuador's public were kept informed and felt that they had some input into the process. Peru's leaders, however, kept their diplomatic strategy and initiatives to themselves. They made little effort to consult outside a narrow official circle. Key elites and the general public in Peru were not kept abreast of developments and had no input into the process.[31] Ecuador's leaders felt that they had no option but to pursue a more inclusive public diplomacy at home, because the government and population alike had long considered the trans-Andean territory to be a defining component of national identity.[32]

Another important difference between the two countries was that Ecuador's military was stronger in terms of domestic political influence and more unified than Peru's. It had built up strength over more than a decade to prepare for an eventual confrontation with Peru. In the 1995 conflict Ecuador's military showed it was ready, successfully keeping Peruvian forces at bay. Because Ecuador's armed forces had the most at stake in any settlement with Peru, top officers needed to be consulted regularly by the diplomats involved in the negotiating process. Peru's military had been caught unprepared in 1995 but had overcome its material differences quickly and wanted the opportunity to get back at Ecuadorean forces. The Peruvian armed forces were also much more dependent on President Fujimori, however, who had over the course of his long presidency made military leadership appointments based on personal loyalty rather than independent capacity.

In short, internal political considerations in Ecuador required that the diplomats check back regularly with key sectors to apprise them of developments and to get their feedback. Over the course of the extended negotiating process, then, many Ecuadoreans were informed in general terms of what was going on and had time to reflect on and reconsider their long-held views. In addition, top-ranking diplomats made public statements from time to time, making it clear that Ecuador could not expect its historic position to win out. When the outside mapping experts named by the guarantors to a special border commission submitted their report in May 1998, one of their key conclusions was that there was no separate watershed other than the Cordillera del Cóndor in the disputed area, thereby undermining Ecuador's historic position. At this critical moment, Ecuadorean authorities disseminated some of the conclusions in the most general terms to prepare the public without violating the proviso forbidding any public declaration of the report. In addition, these same officials began to tout the Treaty of Commerce and Navigation contemplated in the 1942 Rio Protocol as the vehicle by which Ecuador would pursue its Amazonian ambitions, and the country's major newspapers came out in support of the proposed settlement. So when the signing actually took place, most of Ecuador had been prepared for the outcome and accepted it.[33]

This did not happen in Peru. Official information on the diplomatic process was carefully filtered. A government ad campaign in the Peruvian media emphasized the need to "respect the protocol and make it be respected," treating it as an inflexible instrument. This gave the explicit impression that the Peruvian government would be firm and that a territorial adjustment of any kind was totally out of the question. Because the government made no effort to inform the public, even in general terms, most Peruvians were not prepared for the symbolic private property solution that was reached. In addition, most did not understand that the Treaty of Commerce and Navigation was actually referred to in the original Rio Protocol as an instrument that could be negotiated to help implement the terms of the protocol.

As a result, in the disquiet created when the actual terms of the settlement became public in Peru, the political opposition took advantage of the moment to organize a popular protest, which turned violent in some parts of the country, especially in the northern jungle city of Iquitos. While the government quickly quelled the opposition to the agreement, with a number of civilian deaths in the process, to many Peruvians the outcome represented surrender rather than victory.

Peruvian government officials achieved acceptance because they had the domestic political and military power to do so, not because they convinced their citizens that they had achieved almost all that they had sought. Ecuador, even with much more at risk in the negotiations after decades of claims of sovereignty in the Amazon and a population whose national identification rested to a significant degree on these claims, did not experience a violent popular reaction even though the agreement did not satisfy these aspirations.[34] The major reason for such a response was authorities' carefully considered process of consensus building through constant consultation and effective domestic public diplomacy.

Conclusion

The "weight of history" was very much against a peaceful resolution of the Ecuador-Peru conflict. The longer the dispute festered without resolution, the more nettlesome it became. Ecuador's performance in the 1995 small-scale war, combined with the availability of modern weapons, could easily have propelled the two countries into a catastrophic major war.

Fortunately for Ecuadoreans, Peruvians, and everyone in the Western Hemisphere, important changes in the balance of military and diplomatic power as a result of the 1995 hostilities gave a new relevance to the international institutional context within which the conflict was embedded. Leadership, including the diplomatic skill of hemispheric diplomats and the personal courage and foresight of Ecuadorean and Peruvian presidents alike, helped to generate terms that allowed each nation to conclude that resolution at last made more sense than continued conflict. As difficult, drawn out, and problematic as the process was, eventually the diplomacy of cooperation won out over the diplomacy of war.

What could be considered the most important elements that contributed to this felicitous outcome? One is certainly the prior existence of a framework within the international system—the 1942 Treaty of Peace, Friendship, and Boundaries (the Rio Protocol)—that made this dispute, however long-standing, less subject to the system's anarchic characteristics. The problem was that this framework could not be utilized for some thirty-five years due to the shift in position of one of the parties to the treaty, with the result that the dispute returned to its anarchical context, characterized by constant tension and unpredictability. Therefore, the crucial first step toward any possible resolution through cooperation rather than war required a return to the international framework offered by the treaty. Ecuador's decision to do so in early 1995, based on its calculation that the costs of return would be less than the uncertainties of continuing with the status quo, immediately increased the chances for a negotiated settlement.

Another major element was the degree to which the Rio Protocol contained the provisions needed for moving forward toward resolution of the dispute. These included specifics as to how and where the boundary should be drawn, a complementary treaty of commerce and navigation to be negotiated separately, and a multilateral instrument by which four specified friendly countries could provide their good offices to help the parties work out their differences.

The role of the guarantors proved to be another absolutely essential element once the diplomatic process got under way. Unlike the 1950s, this time they responded quickly, worked effectively together, operated at all times with patience and forbearance, made certain that ultimate responsibility for decisions rested with the parties, and had the skills necessary to devise solutions to multiple impasses as they came to the fore. At no point was this more in evidence than in the final stages of the negotiations, when they devised a solution based on what amounted to binding arbitration with a symbolic cession of one square kilometer of territory in the disputed area to Ecuador in permanent usufruct, not sovereignty. In addition, the U.S. guarantor representative, who came to fulfill the role of coordinator among his colleagues through his individual abilities, was able to utilize his standing with the parties to work successfully through the most difficult moments in the process.

The establishment in the disputed area of a multilateral military mission (MOMEP) that included contingents of all of the guarantors as well as the parties as the first major initiative proved to be a critical step. While MOMEP's presence was small and almost symbolic (less than seventy-five personnel permanently deployed), it served as a deterrent to further military confrontations by the parties. By including contingents from both Ecuador and Peru, it also offered a continuing basis for regular contact and confidence building to help defuse the long tradition of hostility and mistrust. Furthermore, since the cost of MOMEP was assumed by the parties to the conflict, they also had a stake in ensuring its effective operation.

Ecuador had always had a greater stake in finding a solution that would satisfy long-held and deeply embedded national aspirations, so the process by which its leaders worked to prepare domestic constituencies for an outcome that would fall short of meeting them was another critical element in resolving the dispute. Officials not only consulted with key elites at each step of the negotiations but also made public statements that expressed adjustments in historic positions and supported alternative options to help set the stage for a settlement that would have public acquiescence, if not support. They succeeded in this effort, in spite of operating within a democratic context that was both fragile and unstable.

Ecuador's pursuit of domestic public diplomacy contrasted markedly with Peru's approach, which consistently kept the details of the negotiating process confined to a small group of officials. Given Peru's stronger initial hand in the context of international law and negotiation dynamics that further strengthened that position over time, the failure to consult or share information about developments at any point was surprising, to say the least. The explanation may lie in a combination of an authoritarian presidential leadership style and an internal democratic process at the time in which the president's supporters had a majority in congress and were amenable to responding as they were asked. Whatever the reasons, the government paid a high price for its hermetic approach when the terms of the final settlement did become public. Even though any objective observer would conclude that Peru gained virtually everything that it had been seeking, violent popular protests rocked the government and spoiled a moment that should have been one of celebration.

Finally, whatever the differences in the ways each country informed its citizens on the state of the negotiations, the roles of the heads of state proved decisive at the final stage of the process. Both President Fujimori and President Mahuad engaged in a remarkable feat of personal diplomacy to break through the resistances of some of the actors, particularly the military and the Peruvian foreign minister, to find a way to a final settlement that made imaginative use of but did not violate the terms of the Rio Protocol. Without their initiative and resolve, it is unlikely that a definitive resolution of the dispute would have been achieved.

Considered separately, none of these elements would have been sufficient in and of itself to provide the basis for a comprehensive settlement. Taken together, however, they worked over time to reinforce the negotiating process and, ultimately, to resolve the conundrum of Latin America's oldest and most enduring border dispute as well as to sustain momentum through the implementation process that followed. This included the placing of the final boundary marker in May 1999, which brought an end to the role of the guarantors under the Rio Protocol after fifty-seven turbulent years.

7 CONCLUSIONS
Lessons Learned, Progress Achieved, and Implications for Other Boundary Disputes

The economic benefits of a resolution of the boundary dispute were clear to both Ecuador and Peru. At the same time, however, some observers felt that such benefits, while positive, were unlikely to have a significant macroeconomic effect on either country.[1] Diplomats on both sides were interested in putting the dispute behind them. Yet the dispute festered for another fifty years after war and a peace treaty ostensibly had produced a settlement. With hindsight, we can describe the Ecuador-Peru boundary dispute as not yet "ripe" for resolution; from our analytic perspective that means that political leaders on both sides would be punished politically if they made the key concessions that would resolve this dispute.

The institutional structure of the democracies created in the late 1970s in both countries was such that it produced a situation in which it was not in any politician's interest to agree on settlement terms. For Ecuador, with no reelection, legislators might have taken a moderately unpopular stance that promised important benefits in the medium term and voted for a settlement. But the economic benefits were likely to be large only for the lightly populated border regions rather than for the nation as a whole. In addition, with Peru insisting on the sanctity of a treaty denounced in Ecuador by democratic and authoritarian leaders alike over decades, no Ecuadorean president who cared about a place in history or legislator hoping to become president or run for other public office in the new democracy could accept a treaty under such terms. For Peru, where reelection was permitted, no legislator or president could accept a resolution adjusting any terms of the 1942 treaty without expecting punishment at the hands of the electorate, given multiple governments' historic position on the border issue.

The stalemate produced in the 1995 war, however, in sharp contrast with Peru's quick military victory in 1981, shifted the power relations between the two countries. This shift altered both the internal and international context within which the institutional constraints on leaders operated. Regionally, the probability of a renewal of hostilities and its development into a major war propelled third parties to commit time, energy, and resources to resolve this dispute. Domestically, Ecuador's ability to stand up to Peru in the battlefield now instilled in both leadership and citizens alike a sense of honor and justice. The potential for war, in turn, threatened both Peruvian president Alberto Fujimori's strategy for strengthening his increasingly authoritarian hold on domestic politics and his neo-liberal economic plan.

The 1995 war, along with the institutional changes that were taking place in Peru, combined to achieve an important goal for Ecuador; they forced Peru to the negotiating table. There, with the active partnership of the guarantor countries, a package of items could be put together that had the potential to turn a zero-sum issue (my territory or yours) into one that was divisible and thus positive-sum.

Ecuador's military performance and Fujimori's willingness and ability to take an unpopular stand made it possible for both sides to accept the guarantors' decision to recognize Peruvian sovereignty over almost all of the territory in dispute while giving Ecuador a symbol of respect in Tiwintza and potentially significant economic benefits as well.[2] Ecuadorean legislators, facing reelection, could now find it in their political interest to accept a resolution that—while far less than the banner in Quito's central plaza proclaiming that "Ecuador is, and has always been, an Amazonian nation"—brought respect and put an old issue behind them as they confronted newer and more pressing domestic political and economic challenges.

Within the ambit of bilateral relations since the 1998 peace agreement, the range of activities which helped to consolidate the peace has significantly increased. Commercial activities have expanded markedly, with Ecuadorean exports to Peru increasing by more than twenty-fold, from just $69 million in 1995 to almost $1.5 billion in 2007 (over half made up of oil and oil products).[3] During the same period, Peruvian exports to Ecuador, while less dramatic in their expansion, still grew over 700 percent, from $47 million to $347 million.[4] Private investment between the two countries also grew significantly, from less than $1 million in 1992 to over $35 million a decade later.[5] These changes have occurred along with a tripling of border area commerce and a five-fold increase in vehicle traffic across the frontier over the past decade.[6]

Another area of significant positive change as the result of the peace treaty is through the Binational Development Plan, which focuses on a variety of infrastructure projects to benefit the almost 5 million residents of the frontier region (1.6 million Ecuadoreans and 3.1 million Peruvians). The projects cover a gamut of basic improvements (such as highway construction, electrification, potable water, sanitation, public health and education facilities) and are financed by a variety of sources (including local and national governments of Peru and Ecuador, over a dozen foreign governments, and a substantial number of NGOs). From 1999 through 2006 total expenditures on such types of projects on the Peruvian side of the border came to almost $1.3 billion, with the total on the Ecuadorean side during the same period reaching just over $700 million.[7] For Peru, the resources expended have financed about 360 separate projects providing benefits to close to one-third of the total frontier area population.[8] On their side of the border, Ecuadorean officials have financed some 530 projects for the benefit of about 340,000 residents of the country's eight frontier provinces.[9]

A third arena of substantial change since the 1998 signing of the peace treaty can be found in the progressive expansion of bilateral meetings by government officials, with regular joint gatherings of cabinet members and key military authorities as well as state visits by both presidents. Symbolizing the degree to which historic barriers have come down was the incident-free 2008 visit to Lima by Quito's mayor, retired general Francisco "Paco" Moncayo, commander of Ecuadorean forces in the 1995 war, during which he characterized bilateral relations as "magnificent."[10] Among other signs of full official normalization are agreements for bilateral military cooperation to demine the frontier over the next five years and for a five-year extension of the Peru-Ecuador Binational Plan for Peace and Development.[11]

Some differences remain, however, as in the delays in utilizing the significant resources promised by international financial institutions (IFIs) due to the complexities of coordinating projects at multiple levels of both governments and to the difficulties of local authorities in meeting the requirement that they provide 20 percent of the financing needed for the projects to go forward.[12] Another difference is to be found in the approaches that the governments of Peru and Ecuador are following in their economic development strategies. Peru is committed to economic liberalization and closer relations with the United States through a Free Trade Agreement, while Ecuador pursues a more statist approach and closer ties with Venezuela. The possibility of a sea boundary conflict is another potential issue, as Peru has taken a different interpretation of the 1950s tripartite accords with Ecuador and Chile for the location of the Peru-Chile line in a submission to the International Court of Justice (ICJ) that could reopen the issue of the Peru-Ecuador line as well.[13] A final potential source of tension is the failure to date either to complete the road from the border to Tiwintza (as agreed in the peace treaty) or to build the docks and storage facilities for Ecuadorean products on the Amazon, part of the complementary 1998 Treaty of Commerce and Navigation.

In spite of such lingering issues and potential areas of bilateral problems, however, the advances achieved in relations between Peru and Ecuador in the decade since the resolution of the border dispute offer clear signs that the problems that divided the two nations for so long have been overcome. Given the progress that has been made in such a short time and the growing interrelationships in both political and economic arenas that have been achieved, it appears likely that any remaining differences will be worked out in constructive and mutually satisfactory ways.

Lessons from the Ecuador-Peru War and Peace

Though Ecuador and Peru resolved their boundary dispute, Tables 1.2 and 1.3 (see Chapter 1) remind us that border issues in Latin America still abound. Eight disputes have been resolved since 2000 (see Table 1.1), most recently Nicaragua/Costa Rica in 2009 and the Peru/Ecuador (maritime boundary) in 2011.[14] In addition, two are at various stages of ICJ processes (Nicaragua/Colombia and Peru/Chile), another one is in an OAS-mediated process (Guatemala/Belize), and there appears to be a new willingness on the part of Chile and Bolivia to address their dispute.[15] In addition, although it was not a border dispute per se, Argentina took Uruguay to the ICJ, alleging a treaty violation dispute over presumed environmental damage from a paper mill along the Paraná River bordering both countries.[16] A decision on April 20, 2010, on balance supported Uruguay.[17]

Even with such positive developments, however, the overall picture is less sanguine. The 2003 ruling by the ICJ to resolve the Honduras/El Salvador dispute in the Gulf of Fonseca inadvertently created a new controversy over the Isla de Conejo. The court's decision on the Nicaragua/ Colombia dispute over the San Andrés Islands in 2006 specifically left maritime boundaries to be worked out by the parties. The need to negotiate sea boundaries has also become important in the Peru/Chile relationship despite the final resolution of their land-based disagreements; it is also interfering with the Bolivia/Chile discussions over Bolivia's claim to sovereign access to the sea. When Bolivia and Chile appeared to be on the verge of a tentative solution in 1977, Peru invoked the clause in the 1879–1883 War of the Pacific settlement that gave it the right to decide on any proposed territorial transfer involving lands which were formally Peruvian (the Department of Tarapacá). Chile could not cede former Bolivian territory without physically dividing the country, so Peru's objections rendered moot any possible territorial transfer.[18] The high price of energy has turned maritime boundaries into an issue much larger than fisheries, as countries scramble to sign contracts with international oil companies to explore for fossil fuels in disputed territorial seas.

Nonboundary disputes have also become more salient over the past few years. Colombia's efforts to resolve its decades-old civil war have created new tensions along both the Ecuadorean and Venezuelan borders. Controversies over Brazilian public and private investment in Ecuador have led to the recall of the Brazilian ambassador. Brazil has also warned Paraguay over the seizure by landless peasants of Brazilian-owned farms. Peru has protested to both Ecuador and Bolivia over their activities promoting leftist organizations involving Peruvian citizens within their territories and in Peru as well.

The vast majority of these disputes, territorial or not, are unlikely to deteriorate into violent conflict; but disturbing signs suggest that we should not believe that Latin America has reached a state in which the use of force as an instrument of statecraft has been rendered illegitimate or null. The willingness to appeal verbally to force, to demonstrate one's ability to mobilize force, and even to employ force short of war has not disappeared in the region.[19] For example, when Paraguay's government insisted on a new accord that would increase the price of electricity that it sold to Brazil from the binational Itaipú Dam installations and failed to respond quickly to invasion of Brazilian-owned farms by landless Paraguayan peasants, Brazil's government mobilized troops to the border. Believing that it was being unduly pressured by its powerful neighbor, the Fernando Lugo government complained to the OAS Permanent Assembly about Brazil's behavior.[20]

In this context of boundary and other disputes, a long-postponed modernization of Latin American militaries fuels uncertainty and suspicion. Honduran purchases of military aircraft increase tension with Nicaragua, which refuses to destroy more SAM-7 missiles unless Honduras and El Salvador rid themselves of military aircraft.[21] Chile's purchase of F-16C fighter jets, state-of-the-art Leopard tanks, and Humvees are reported to be "of great concern to Peru, Bolivia, and Argentina."[22] Venezuela's military modernization project has accelerated in the last two years, with US$4 billion spent on fighter jets, attack helicopters, and 100,000 Kalashnikov assault rifles and ammunition. There are also plans to build a Kalashnikov production facility. This, combined with claims that the specific caliber of ammunition used in these Kalashnikovs is also usable in the older models employed by Fuerzas Armadas Revolucionarias de Colombia (FARC, Revolutionary Armed Forces of Colombia), raises fears about leakage from the Venezuelan army to guerrilla groups in the region.[23] Venezuela's military assistance to Bolivia is nontransparent and confusing to Bolivia's neighbors,[24] especially after Hugo Chávez announced that his military would come to the aid of the Evo Morales government when the autonomy movement among the eastern departments showed signs of leading to civil war.[25]

The U.S. agreement with Colombia over access to seven military bases also contributes to concern. The two countries insist that the U.S. military presence will continue the fight against the drug trade within Colombia. Yet U.S. support for Colombia's cross-border incursion into Ecuador makes many worry that a threshold in the war against drugs and terror has been crossed. Proclamations by the U.S. Department of Defense before the U.S. Congress about achieving increased operational capabilities in South America once the bases are upgraded are disconcerting to Colombia's neighbors.[26]

The military balance is thus becoming more ambiguous at a time when the institutional constraints on political leaders are making them more accountable to their citizens. The Ecuador-Peru case demonstrates that democratic citizens can favor the use of military force when they believe the costs are low and national interests are at stake. It is disconcerting that 82 percent of Colombians supported President Alvaro Uribe Vélez's decision to use army units to pursue the FARC across the border into Ecuador,[27] in clear violation of international law.

Unfortunately, a number of Latin American leaders of countries with outstanding boundary and other disputes fall into the category of risk

Table 7.1. Militarized Interstate Disputes, 2005–2009

Date	Countries	MID score
January 2005	Venezuela, Colombia	3
October–November 2005	Costa Rica, Nicaragua	2
February 2006	Colombia, Ecuador	3
April 2007	Nicaragua, Costa Rica	4
May 2007	Argentina, Uruguay	3
July 2007	Colombia, Nicaragua	3
February 2008	Colombia, Nicaragua	3
March 2008	Colombia, Ecuador	4
March 2008	Colombia, Venezuela	3
2009	Venezuela, Colombia	3

Sources: "Venezuela Breaks Ties with Colombia after Rebel Spokesman Kidnapped from Caracas," *NotiSur*, January 28, 2005; "Río San Juan: From Nicaragua and Costa Rica to The Hague," *NotiCen*, November 3, 2005; Gabriel Marcella, *War without Borders: The Colombia-Ecuador Crisis of 2008*; "Security Update," *Latinnews Daily* (Intelligence Research Ltd), April 25, 2007, http://www.latinnews.com (accessed May 22, 2007); "Interstate Conflict in Latin America: A Thing of the Past?" *Latin American Special Reports*, April 2007, http://www.latinnews.com (accessed May 22, 2007); Simon Romero, "Talk of Independence in a Place Claimed by 2 Nations," *New York Times*, February 1, 2008; *Keesing's Record of World Events*, 54 (March 2008), Colombia, 48456; Brian Ellsworth, "Venezuela Mobilizes Forces to Colombia Border," *Reuters*, March 5, 2008, http://www.reuters.com/article/2008/03/05/us-venezuela-colombia-idUSN0227633020080305 (accessed June 24, 2011); Guy Adams, "Tensions Grow as Chavez Masses Troops on Border," *Independent*, November 10, 2009, http://www.independent.co.uk/news/world/americas/tensions-grow-as-chavez-masses-troops-on-border-1817728.html (accessed April 18, 2010).

takers, but without being innovators. Outside of Bolivia and Brazil, both the Left (Venezuela, Ecuador, Argentina, and Nicaragua) and the Right (Colombia, Peru, and Mexico) are using traditional approaches to address today's challenges. That means that they are not working to find new ways to solve problems, which arise precisely because the old approaches have proven incapable of resolving them. In interstate disputes, such "old ways" of addressing issues include militarizing them.

The relevance of the Ecuador-Peru war and subsequent peace has been thrust into the Western Hemisphere by the March 2008 Colombian attack on a guerrilla camp across the border in Ecuador after Colombia claimed that the guerrillas had opened fire. Ecuador denounced the violation of its territory and mobilized its military. Venezuela jumped into the fray, breaking relations with Colombia and ordering troops to the border while warning Colombia that violation of its territory would mean war.

Nicaragua also suspended relations with Colombia. For a few days Latin Americans and many observers outside the region held their breath, until at last reconciliation was brokered. Although Colombia's relations with both Ecuador and Venezuela soon deteriorated again, President Juan Manuel Santos Calderón (2010–2014) has taken several significant steps to reestablish normal ties.

Militarization occurs even in conditions of great power disparities and the presence of international institutions. In 2009 the ICJ resolved a disagreement about the rights of Costa Rica in the San Juan River, which defines the border with Nicaragua; but in October 2010 Costa Rica discovered that the Nicaraguans were clearing a channel on a marshland in a disputed section of the Isla Calero. Costa Rica has no army, but Nicaragua sent dozens of troops to the area in a clear effort to intimidate the Costa Ricans when they complained. The OAS attempted to mediate the dispute, but Nicaraguan president Daniel Ortega claimed that the organization had no jurisdiction over territorial disputes. Costa Rica filed a claim with the ICJ in November 2010. In March 2011 the court issued an order for the two parties to remove their personnel (Costa Rica had dispatched police to the area) while the case proceeds.[28]

Unfortunately, as Table 7.1 indicates, in the period from 2005 to 2009 military force has been threatened, mobilized, and even used ten times, involving seven Latin American countries, not just Colombia, Venezuela, and Ecuador.

The Role of Third Parties

As our case study demonstrates, third parties can play a fundamental and positive role in helping countries to find ways to resolve their disputes. An offer by third parties of help in resolving interstate disputes may have unintended consequences, however, because of "moral hazard." This situation arises when an insurance policy against some risky behavior lowers that risk sufficiently so that the covered party decides that the benefits of such behavior now outweigh the costs. As a result, the insurance winds up stimulating risky behavior. Ignoring "moral hazard" possibilities may encourage hard-line positions by weaker parties in the dispute, including violence, in the hope that an interested hemispheric community might increase pressure on a rival to settle. Although the hemisphere rejoices that Ecuador and Peru have settled their dispute, one should remember that it took a small war in 1995 and the threat of a large one in 1998 to help convince the parties to settle. By guaranteeing that conquest will not be recognized and that escalation into a costly and long war will be unlikely, the presence of the OAS in the 1981 conflict and the guarantor country mechanism established in the Rio Protocol may have helped convince Ecuador to engage in the adventurous behavior that developed into the short war of 1995.[29]

Insurance companies deal with the problem of moral hazard by including clauses in their contracts that either discourage such behavior directly (for example, charging lower fees for nonsmokers) or free the company from covering losses due to certain behavior. The hemispheric community will need to pay as much attention to providing these disincentives as to promoting resolutions themselves. We end this book by offering some suggestions for distinguishing between those boundary disputes which might be ripe for settlement and those which might be moving toward a dramatic escalation of tensions.

We need to begin with a prudent understanding of the benefits of dispute resolution rather than assume that dispute resolution per se should drive policy. The focus should be on the defusing or resolution of those specific disputes that are ripe for militarization or may be *significantly* undermining democracy, economic cooperation, or social welfare due to particular national situations. Boundary disputes that have minimal if any impact on bilateral economic, political, or social relations, such as those between Uruguay and Brazil, should be left to find their own time for "ripening" toward resolution.

Prudence does not necessarily imply waiting for a conflict to militarize before focusing attention upon it. Yet "early warning systems" are notoriously unreliable. The challenge is that many points of potential conflict exist and many disagreements begin to become tense yet never develop into the type of crises that bring third-party states and international institutions to invest resources into attempts to defuse or resolve the dispute. Early indicators of success can also be misleading; promising bilateral negotiations between Ecuador and Peru (1990–1995), Bolivia and Chile (1977), and Chile and Argentina (1972–1977) collapsed and produced war scares among two of the dyads as well a short war in the third. The wealth of false positive signals can produce multiple problems by well-meaning third parties. Hemispheric institutions or individual parties may become involved, encouraging the disputants to expect their aid in resolving the issue; however, when the issue turns out not to escalate, the outsiders back off, leaving at least one of the disputants to feel abandoned; Ecuador and Bolivia are such examples.

Given the problems of moral hazard and early warning, perhaps the best approach for third parties is the traditional one: to wait for the disputants themselves to ask for mediation or arbitration. Forcing the disputants to the bargaining table is unlikely to prepare the domestic context for the bargains that have to be struck. Once both parties make such a request, the international institutional structure is well set up to respond to requests for mediation or arbitration, at the global (for example, ICJ), regional (for example, OAS), or even subregional (for example, Mercado Común del Sur [Mercosur, Southern Common Market]) levels.

Conclusion

Our conclusion about the optimal situation in which a settlement is most likely considers three elements. First, domestic institutional structures will emphasize accountability over representation. Second, the costs of using force will be clear and considerable for at least one of the parties to the conflict. Third, leadership will be innovative and risk acceptant. Given this context, international aid can help tip the balance; but if any of the three elements are absent, third-party involvement will be insufficient.

At times nations may have goals that cannot be reached or whose achievement would undermine efforts to reach more important goals. Because international politics takes place in a context of strategic interaction and resources are distributed unequally, "sovereignty" cannot mean that a nation can do whatever it pleases but rather that the national government has the ability to choose which response serves its interests best, given that international context. Leadership entails recognizing these instances and devising policies to settle the issue or relegate it to the back burner until such time as the nation can accept what it once regarded as an unfavorable solution.

Because the relevant decision maker is not a bureaucrat or simply an appointed administrator, settlement is not a matter of political will but of politics. In foreign policy matters, the head of government (whether president, prime minister, or dictator) must authorize decisions that commit the government either to continue or to resolve a boundary dispute. Therefore, the costs and benefits to politicians' constituencies are key items upon which to focus.

Policy choices are best understood from this perspective. While domestic policy choices probably matter more for elections,[30] in cases where foreign policy is disputatious it must matter to important domestic actors (or else there would have been a settlement). Leaders therefore will approach boundary disputes with a sense of the domestic political limits and opportunities created by the coalition of political forces that support them in office.

Even periods of domestic strife offer opportunities for taking positive actions in foreign affairs, rather than simply emphasizing threats from outsiders. Taking a major step toward resolving a contentious issue can benefit a leader at these times by illustrating the ability to rise above "partisan" politics and act for the good of the nation.

Our institutional analysis suggests that the conditions under which the incentive structure of politics can support a resolution are twofold. First, a president needs to think like a political entrepreneur who can see the benefits of resolving the disputes for the head of state's major programs. These programs are likely to be oriented around the social and economic development of the country. Any reformist president will have to fight important political battles for the government's major programs. This brings us to the second requirement. Economic reforms need to generate payoffs relatively quickly. (The international community can contribute to the peace process with support for economic development, as it did with Ecuador and Peru.) Once those programs begin to produce, the head of state may be able to use these successes to bring public opinion and the army around to the position that settling the international issues is a contribution to the continued development of the nation.

Because politics is the art of the possible, leaders are rational actors, and individuals and groups can be expected to respond to appropriate incentives, boundary disputes are not "destined" to remain a part of the hemispheric landscape. The challenge for the international community is to avoid becoming overzealous on the issue. By demonstrating a cautious willingness to help when asked and having the patience to work to develop "ripeness" for resolution, the hemispheric community can best contribute to advancing the day when the Americas can turn their undivided attention to political stability, economic prosperity, and social welfare for all.

APPENDIX A
Effective Number of Parties

Octavio Amorim Neto

The nearly consensual solution to the problem of determining the effective number of parties is to employ the Laakso and Taagepera effective number of legislative (or electoral) parties.[1] The Laakso and Taagepera index has the formula $N = 1/\Sigma x_i^2$, where x_i is the percent of seats or votes held by the ith party. By squaring the legislative or electoral size of parties, this index overestimates the weight of larger parties and underestimates the size of smaller ones, thus providing a nonarbitrary criterion to identify the number of relevant parties either in elections or in legislatures. It varies from 1 (when all seats or votes are held by one party) to the number of seats available in a given legislature (when each seat belongs to a different party).

Yet there is a problem in applying N to Ecuador. The index assumes that all seats are held political parties. However, it turns out that the Ecuadorean legislature sometimes has an impressive number of independents. Fortunately, Taagepera has recently proposed an index to calculate the effective number of parties in the presence of independents.[2] The new index proposed by Taagepera uses a formulation of N that avoids detailed calculation of fractional shares:

$$N = P^2/\Sigma_i P_i^2$$

where P_i stands for the number (rather than fractional share) of seats or votes for the ith party, and P is the total number of seats or valid votes. If there is a residual of R votes or seats lumped as "Other," the expression becomes

$$NTC = P^2/[f(R) + \Sigma P_i^2]$$

where $f(R)$ is a function of R to be estimated and the summation extends only over the individually specified parties. One can determine the smallest and largest values that $f(R)$ could possibly take for a given R and then calculate the corresponding N. This determines the possible error range for N, and in the absence of other information the average of the extremes might be used.

The largest value of N is obtained when every item in R (votes or seats) belongs to a different party, so that the sum of their squares is $f(R) = 1^2 + 1^2 + \ldots = R$. In most cases it is very small compared to the summation of known P_i^2, so that we can approximate $f(R) = R$ with $f(R) = 0$.

To calculate the effective number of parties in Ecuador, we will average N and NIC. As for NLC, P is the total number of legislative seats, and P_i is the number of seats held by the ith party. Independent legislators have the function of the residual in NIC. I will assume that each independent represents a different party exactly to obtain the largest fractionalization of the legislature. As for N, we will consider all independents as though they formed one party. This will allow me to calculate the smallest fragmentation of Ecuadorean legislatures. The final number of effective parties is the average of the values of N and NIC for each legislature.

Example: suppose a legislature with 20 seats. Party A has 8 seats, party B holds 6, and independent legislators hold 6 seats. The effective number of parties of this legislature is calculated as follows:

$NIC = 20^2/1^2 + 1^2 + 1^2 + 1^2 + 1^2 + 1^2 + 8^2 + 6^2 = 400/6 + 64 + 36 = 400/106 = 3.8$

$N = 1/0.4^2 + 0.3^2 + 0.3^2 = 2.9$

Effective Number of Parties $= (NIC + N)/2 = 3.35$

APPENDIX B
Ecuadorean Attitudes toward Relations with Peru
(November 1992)

1. Some people say that Peruvians and Ecuadoreans are very similar people. Others say that there is a great difference between a Peruvian and an Ecuadorean. With whom are you more in agreement?

    34% very similar; 60% great difference

2. Do you believe that it would be convenient to open commerce between Ecuador and Peru completely, as was done with Colombia?

    55% yes; 39% no

3. Do you believe that a solution to the border problem could contribute to the economic development of both countries?

    79% yes; 15% no

4. Do you see Peru as a friendly country or an enemy?

    39% friendly; 49% enemy

Source: Jaime Durán Barga, "Actitud de los ecuatorianos frente al Perú: Estudio de opinión pública."

APPENDIX C
Polling Data on Border Issues (1994–1996)

Figures given in percentages; numbers do not necessarily add to 100 percent due to rounding.

Do you believe that the government has accepted the Rio Protocol or does it continue to consider it null? (February 1994 and January 1996)

Year	No Answer	Accepted	Not Accepted
1994			
Quito	7.3	23.3	69.5
Guayaquil	5.3	14.0	80.0
1996			
Quito	12.0	34.5	53.5
Guayaquil	10.0	32.8	57.3

Do you agree or disagree with the position of the government on the Rio Protocol? (February 1994)

Quito

Of those who did not answer, 54.5 percent did not answer the question about whether the government had accepted the protocol or not; 18.2 percent believed that it had accepted the protocol; and 27.3 percent believed that it had not.

Of those who agreed with the government, 5.8 percent did not answer the question about whether the government had accepted the protocol or not; 24.0 percent believed that it had accepted the protocol; and 70.1 percent believed that it had not.

Of those who disagreed with the government, 6.2 percent did not answer the question about whether the government had accepted the protocol or not; 21.0 percent believed that it had accepted the protocol; and 72.8 percent believed that it had not.

Guayaquil

Of those who did not answer, 37.5 did not answer the question about whether the government had accepted the protocol or not; 12.5 percent believed that it had accepted the protocol; and 50.0 percent believed that it had not.

Of those who agreed with the government, 5.3 percent did not answer the question about whether the government had accepted the protocol or not; 14.2 percent believed that it had accepted the protocol; and 80.5 percent believed that it had not.

Of those who disagreed with the government, 1.4 percent did not answer the question about whether the government had accepted the protocol or not; 17.6 percent believed that it had accepted the protocol; and 81.1 percent believed that it had not.

Was the declaration of peace signed by Peru and Ecuador in Brazil beneficial or prejudicial for the country? (February 1995)

City	No Answer	Beneficial	Prejudicial
Quito	9.8	47.5	42.3
Guayaquil	5.8	55.0	39.0

Do you believe that the international observers who are in the war zone can stop it? (February 1995)

City	No Answer	Yes	No
Quito	3.0	40.0	57.0
Guayaquil	4.0	44.8	51.3

Would you be in favor or opposed to the convocation of a referendum (consulta popular) to reach a definitive settlement to the territorial conflict with Peru? (February 1995)

City	No Answer	In Favor	Opposed	Indifferent
Quito	1.5	52.0	45.0	0.8
Guayaquil	2.0	65.0	33.0	0.0

Do you believe that after the last armed conflict Ecuador can still attain a sovereign outlet to the Amazon River? (February 1995)

City	No Answer	Yes	No
Quito	4.0	75.3	20.8
Guayaquil	4.3	83.5	12.3

Do you believe that it is possible for Ecuador to recover all the territory lost in 1941? (February 1995)

City	No Answer	Yes	No
Quito	1.0	27.5	71.5
Guayaquil	2.0	32.8	65.3

If the Peruvians take Tiwintza, what should Ecuador do: undertake a new offensive to recapture it no matter what the cost or continue negotiating? (February 1995)

City	No Answer	Offensive	Negotiate
Quito	4.0	65.3	30.8
Guayaquil	1.5	74.5	24.0

Of the Peruvian candidates for president, who is less of an enemy of Ecuador: Javier Pérez de Cuellar or Alberto Fujimori? (February 1995)

City	No Answer	De Cuellar	Fujimori	Both	Neither
Quito	5.0	31.0	13.8	26.0	24.3
Guayaquil	2.3	69.3	7.8	5.8	15.0

Is the Rio Protocol null or valid? (January 1996)

City	No Answer	Null	Valid
Quito	5.3	53.3	41.5
Guayaquil	5.0	48.3	46.8

There are people who say that Ecuador should recover the territory which it lost in the 1941 war and that it should be done no matter what the cost. (January 1996)

City	No Answer	Agree	Disagree	Indifferent
Quito	1.3	44.3	52.8	1.8
Guayaquil	0.0	56.5	42.0	1.5

Do you believe that it is necessary to delimit the border (cerrar la frontera) according to the Rio Protocol in order to end the territorial problem with Peru? (January 1996)

City	No Answer	Yes	No
Quito	4.0	51.8	44.3
Guayaquil	2.0	47.8	50.3

Who do you think won the war between Ecuador and Peru last year? (January 1996)

City	No Answer	Ecuador	Peru	Neither
Quito	5.0	55.0	4.5	35.5
Guayaquil	3.3	74.3	2.8	19.8

Which of the two countries was more affected economically by the war? (January 1996)

City	No Answer	Ecuador	Peru	Both	Neither
Quito	2.8	53.0	18.0	26.0	0.3
Guayaquil	1.8	38.5	38.5	21.0	0.3

Do you believe that it is possible that armed confrontations will recur in the near future? (January 1996)

City	No Answer	Yes	No
Quito	5.3	78.5	16.3
Guayaquil	5.3	80.8	14.0

If a new conflict develops, who do you think will win this time: Peru or Ecuador? (January 1996)

City	No Answer	Ecuador	Peru	Both	Neither
Quito	18.8	39.5	28.5	4.5	8.8
Guayaquil	16.5	52.0	12.0	8.5	11.0

Source: Archives, Informe Confidencial (Quito) for February 9, 1994, February 25, 1995, and January 6, 1996.

NOTES

Chapter 1

1. Published accounts of total casualties on both sides noted between 200 and 300. Interviews with Peruvian and Ecuadorean military authorities put the figures much higher, however, between 1,000 and 1,500. Adrián Bonilla, "Proceso político e intereses nacionales en el conflicto Ecuador-Perú"; David R. Mares, "Deterrence Bargaining in Ecuador and Peru's Enduring Rivalry: Designing Strategies around Military Weakness." David Scott Palmer, "The Search for Conflict Resolution: The Guarantors and the Peace Process in the Ecuador-Peru Dispute," cites an Ecuadorean diplomat who estimated that there were as many as 4,500 casualties in the conflict (24, 39).

2. For example, the United States identified a number of roles for Latin American militaries in the post–Cold War era, but none included traditional defense against a neighboring country. United States, Department of Defense, *United States Security Strategy for the Americas*. See also Miguel Angel Centeno, *Blood and Debt: War and the Nation State in Latin America*; Brian Loveman, *For la Patria: Politics and the Armed Forces in Latin America*; and David Pion-Berlin and Harold A. Trinkunas, "Attention Deficits: Why Politicians Ignore Defense Policy in Latin America."

3. For example, Clifford E. Griffin, *Power Relations and Conflict Neutralization in Latin America*.

4. The classic formulation is Edward D. Mansfield and Jack Snyder, "Democratization and War." For an application to Latin America, see Carlos Escudé and Andrés Fontana, "Argentina's Security Policies: Their Rationale and Regional Context," 54; and Francisco Rojas Aravena, "Transition and Civil-Military Relations in Chile," 81.

5. David R. Mares, "Boundary Disputes in the Western Hemisphere," 45–48.

6. David R. Mares, *Violent Peace: Militarized Interstate Bargaining in Latin America*, 37.

7. Central Intelligence Agency (CIA), *The World Factbook 2010*; U.S. Department of Defense, Maritime Claims Reference Manual, June 2008; International Boundary Research Unit.

8. Beth A. Simmons, "Trade and Territorial Conflict in Latin America: International Borders as Institutions," 271.

9. Natasha Niebieskikwiat, "Hielos continentales: Reclamo de Chile por los mapas argentinos."

10. "Island Dispute Sours Relations between El Salvador and Honduras."

11. Simmons, "Trade and Territorial Conflict," 286, n. 33.

12. Leandro Area and Elke Nieschulz de Stockhausen, *El Golfo de Venezuela: Documentación y cronología, Vol. II* (1981–1989), 64–87; Simmons, "Trade and Territorial Conflict," 282–283, missed this major incident in her analysis.

13. United States, Department of Defense, *Maritime Claims Reference Manual*, 2–115.

14. Juan Forero, "Another Bump in a Rocky Road for Colombia and Venezuela," A12.

15. Douglass C. North, *Institutions, Institutional Change and Economic Performance*; Sven Steinmo, Kathleen Thelen, and Frank Longstreth, eds., *Structuring Politics: Historical Institutionalism in Comparative Analysis*.

16. The literature on war and democracy emphasizes the factors of limited government, accountability, transparency, and competitiveness in defining democracy. While the literature about what constitutes democracy is vast, that debate is marginal to our analysis here.

17. Edward D. Mansfield and Jack Snyder, "Democratization and the Danger of War"; Jack Snyder, *Myths of Empire*.

18. The concept of "delegative democracy" (the capacity of elected heads of state to ensure executive branch dominance through various mechanisms) was first developed by Guillermo O'Donnell. See his "Delegative Democracy."

19. Brian Barry, *Sociologists, Economists & Democracy*, 99–100.

20. Febres Cordero did return to be mayor of Ecuador's major city, Guayaquil.

21. Scott Mainwaring and Timothy R. Scully, eds., *Building Democratic Institutions: Party Systems in Latin America*, 1. The fourth category is implicit, as the authors explicitly discuss only the three categories in which party systems exist.

22. Margaret G. Hermann, Charles F. Hermann, and Joe D. Hagan, "How Decision Units Shape Foreign Policy Behavior"; Jonathan Bendor and Thomas H. Hammond, "Rethinking Allison's Models."

23. Cf. Avant D. Deborah, *Political Institutions and Military Change: Lessons from Peripheral Wars*; and Peter D. Feaver, *Armed Servants: Agency, Oversight, and Civil-Military Relations*.

24. Mares, *Violent Peace*, 132–159.

25. J. Lloyd Mecham, *The United States and Inter-American Security, 1889–1960*.

26. See http://www.cubaminrex.cu/english/OAS/Articles/History/inicio.html (accessed June 17, 2011).

27. Almost one-third of all regional trading agreements between 1948 and 1994 occurred in the brief period 1990 to 1994. Jeffrey A. Frankel, *Regional Trading Blocs in the World Economic System*, 4–12. See also W. Ladd Hollist and Daniel L. Nelson, "Taking Stock of American Bonds: Approaches to Explaining Cooperation in the Western Hemisphere."

28. T. V. Paul, *Asymmetric Conflicts: War Initiation by Weaker Powers*; Paul K. Huth, *Extended Deterrence and the Prevention of War*.

29. Alexander L. George and Richard Smoke, *Deterrence in American Foreign Policy*; John J. Mearsheimer, *Conventional Deterrence*; Alan Alexandroff and Richard Rosecrance, "Deterrence in 1939"; Paul, *Asymmetric Conflicts*.

30. David A. Baldwin, *Economic Statecraft*; I. M. Destler and John Odell, assisted by Kimberly Ann Elliott, *Anti-Protection: Changing Forces in United States Trade Politics*; Lisa L. Martin, *Coercive Cooperation: Exploring Multilateral Economic Sanctions*.

31. Thomas P. Anderson, *The War of the Dispossessed: Honduras and El Salvador, 1969*; Mark Rosenberg et al., *Honduras: Pieza clave de la política de Estados Unidos en Centro América*.

32. Peter Gourevitch, *Politics in Hard Times*; Valerie Bunce, *Do New Leaders Make a Difference?*

Chapter 2

1. This brief historical summary is based largely upon Julio Tobar Donoso and Alfredo Luna Tobar, *Derecho territorial ecuatoriana*; Gustavo Pons Muzzo, *Estudio histórico sobre el Protocolo de Rio de Janeiro*; Carlos E. Scheggia Flores, *Origen del pueblo ecuatoriano y sus infundadas pretensiones amazónicas*; Gobierno de Ecuador, Ministerio de Relaciones Exteriores, *Hacia la solución del problema territorial con el Perú: Libro blanco*; Bryce Wood, *The United States and Latin American Wars, 1932–42*.

2. Robert N. Burr, *By Reason or Force*, 44–45, 80–88, 146–147.

3. Frank MacDonald Spindler, *Nineteenth Century Ecuador: A Historical Introduction*, 140–142, 189.

4. *Foreign Relations of the United States (FRUS,* 1910), 439.

5. Spindler, *Nineteenth Century Ecuador*, 30; U.S. and Argentine diplomats also believed that Chile had significant influence in Quito. *FRUS* (1910), 492–493.

6. Ironically, the contemplated settlement in 1910, while depriving Ecuador of vast territory, provided for Ecuadorean access to the Marañón River, which the treaty following the 1941 war would not.

7. *FRUS* (1910), 171–183.

8. Military conflicts between Peru and Ecuador occurred in 1910, 1911, 1912–1913, 1914–1916, 1917–1918, 1932, 1934–1936, 1937, 1938, 1939–1942, 1942, 1943, 1950, 1951, 1953, 1954, 1955, 1956, 1960, 1977–1978, 1981, 1983, 1984, 1985 (twice), 1988, 1989, 1991, 1993, 1994, 1995 (twice), and 1998. Militarized Interstate Dispute (MID) database; Joseph E. Loftus, *Latin American Defense Expenditures, 1938–1965* (Santa Monica: RAND, 1968), 27–29; "No ha habido agresión," *Hoy*, December 30, 1995;

"Peru and Ecuador Hold Fresh Talks," *Financial Times*, September 8, 1998, 9; Carlos E. Scheggia Flores, *Origen del pueblo ecuatoriano y sus infundadas pretensiones amazónicas*, 61; Gobierno de Ecuador, Ministerio de Relaciones Exteriores, *Hacia la solución del problema territorial con el Perú: Libro blanco*, 194–195.

9. Whitney T. Perkins, *Constraint of Empire*.

10. The payoff to Colombia was settled borders with Ecuador and Peru (although they had to fight a war in 1932 to ensure the borders), leaving the country with only disputes with Nicaragua and Venezuela (both of which continue to flare up). Wood, *The United States and Latin American Wars, 1932–42*, 169–172; Ronald Bruce St. John, "The Boundary between Ecuador and Peru," 12; Frederick B. Pike, *The United States and the Andean Republics*, 203–204.

11. Wood, *The United States and Latin American Wars, 1932–42*, 69–251; Pike, *The United States and the Andean Republics*, 266–269.

12. Anita Isaacs, *Military Rule and Transition in Ecuador, 1972–92*, 1–3; Víctor Villanueva, *100 años del ejército peruano: Frustraciones y cambios*, 100–107. The U.S. military attaché in Lima rated Peru's combat efficiency as significantly better than Ecuador's. Daniel M. Masterson, *Militarism and Politics in Latin America: Peru from Sánchez Cerro to Sendero Luminoso*, 65–70; Wood, *The United States and Latin American Wars, 1932–42*, 268–269.

13. Although Peruvian president Manuel Prado was opposed to a war with Ecuador, the commander of the northern army insisted upon attacking the Ecuadorean forces. Geoffrey Bertram, "Peru 1930–60," in Leslie Bethel, ed., *The Cambridge History of Latin America*, 8:423; Wood, *The United States and Latin American Wars, 1932–42*, 255–331.

14. The full official English translation of the protocol may be found in William L. Krieg, *Ecuadorean-Peruvian Rivalry in the Upper Amazon, Enlarged to Include the Paquisha Incident (1981)*, Appendix 1.

15. See the comments by the Ecuadorean negotiator in Tobar Donoso and Luna Tobar, *Derecho territorial ecuatoriana*, 212–222; and by Gen. José W. Gallardo Román (commander in chief of the army), "Comentario militar," 34–35 (Gallardo was defense minister during the 1995 war). Captain Altamirano's book begins with a discussion of Incan expansionism. See also Lt. Gen. Frank Vargas Pazzos (ret.), *Tiwintza: Toda la verdad* (Vargas Pazzos was leader of one of the political parties in congress upon which President Bucaram depended); and Wood, *The United States and Latin American Wars, 1932–42*, 326–330.

16. Drawing from Krieg, *Ecuadorean-Peruvian Rivalry in the Upper Amazon*, and others, Gabriel Marcella notes that "[t]he signing of the Protocol verified the Status Quo Line of 1936 [Act of Lima, see the historical time line in the main text] signed . . . by Ecuador and Peru, minus the [net] loss to Ecuador of only 5,392 square miles" (*War and Peace in the Amazon: Strategic Implications for the United States and Latin America of the 1995 Ecuador-Peru War*, 6). Wood, however, points out that the status quo was to be maintained "until the termination of the discussions in Washington" (*The United States and Latin American Wars, 1932–42*, 259–260). On the García-Herrera Treaty, which would have cost Peru almost 120,000 square miles (310,000 square kilometers), see St. John, "The Boundary between Ecuador and Peru," 9–11.

17. Dispatch 881, from Quito, August 26, 1937. In *Foreign Relations of the United States (FRUS, 1937)*, 54–55, as cited in Wood, *The United States and Latin American Wars, 1932–42*, 260.

18. The result of Peru and Ecuador's submission to the guarantors of differences that arose in demarcating the border under article 8 of the protocol. República del Perú, Ministerio de Relaciones Exteriores, *Frontera peruano-ecuatoriana: El laudo arbitral de Braz Dias de Aguiar—Reportorio documental*. Details discussed in Krieg, *Ecuadorean-Peruvian Rivalry in the Upper Amazon*, 128–132.

19. The international legal principle of "as you possess, so you will possess." As applied to Latin America, the administrative frontiers of the Spanish Empire became the international frontiers of the newly independent countries. There are two schools of thought related to this principle, however: de jure (the new boundaries should follow the lines on the map of the colonial boundaries, whether or not the territories encompassed had been occupied); and de facto (the new boundaries should be based on physical possession of the territory within colonial boundaries). Ecuador interpreted this principle de jure; therefore Ecuador held that its international boundary followed the administrative borders of the Viceroyalty of Nueva Granada, of which it was a part (and which included a large portion of present-day northeastern Peru). See Tobar Donoso and Alfredo Luna Tobar, *Derecho territorial ecuatoriana*, especially 51–59. Also see a more general discussion in Steven R. Ratner, "Drawing a Better Line: *Uti Possidetis* and the Borders of New States."

20. "Protocol of Peace, Friendship, and Boundaries between Peru and Ecuador" (official English translation), articles 5, 6, 7, 8, and 9, in Krieg, *Ecuadorean-Peruvian Rivalry in the Upper Amazon*, Appendix 1.

21. Krieg, *Ecuadorean-Peruvian Rivalry in the Upper Amazon*, 128–132; República del Perú, Ministerio de Relaciones Exteriores, *Frontera peruano-ecuatoriana*, especially 123–180.

22. Krieg, *Ecuadorean-Peruvian Rivalry in the Upper Amazon*, 132–137. Also summarized in Ronald Bruce St. John, "Conflict in the Cordillera del Cóndor: The Ecuador-Peru Dispute," 79–80.

23. From a ten-page commentary based on U.S. government documents compiled and made available to Palmer by a Peruvian diplomat and student of the problem who prefers not to be named.

24. George McBride, "Ecuador-Peru Boundary Settlement." There is also a report on the trip by the Joint Boundary Commission: Capt. Luis F. Montezuma (Peru) and Lt. Gustavo Proaño (Ecuador), "Informe itinerario #1." A copy of this report was provided to Palmer by William Krieg.

25. McBride's 1949 report speculates upon the meaning of these delays in a manner closely anticipating subsequent Ecuadorean policy. William Krieg, however, believes that Ecuador was genuinely surprised by the map anomaly and speculates that the October 27, 1943, field report was lost in the Ecuadorean military bureaucracy (Palmer interview, September 21, 1996). His view is supported by a U.S. government specialist on the topic, who notes: "Even if both sides knew how far up the Cenepa went, Dias de Aguiar apparently did not know. If he had, he would not have repeated the error of the protocol [in his report]" (Palmer interview with Luigi Einaudi, January 29, 1997).

26. Pons Muzzo, *Estudio histórico sobre el Protocolo de Rio de Janeiro*, 277–278.

27. República del Ecuador, Ministerio de Relaciones Exteriores, *Hacia la solución*, 78–79; Fernando Bustamante, "Ecuador: Putting an End to Ghosts of the Past?"

28. William P. Avery, "Origins and Consequences of the Border Dispute between Ecuador and Peru," 74.

29. Scheggia Flores, *Origen del pueblo ecuatoriano y sus infundadas pretensiones amazónicas*, 71–73.

30. Luis Carrera de la Torre, *El proyecto binacional Puyango Tumbes*, 73–111.

31. Avery, "Origins and Consequences of the Border Dispute between Ecuador and Peru," 69.

32. Stephen M. Gorman, "Geopolitics and Peruvian Foreign Policy," 71, 83–84.

33. Krieg, *Ecuadorean-Peruvian Rivalry in the Upper Amazon*, Epilogue, 266–335.

34. David Scott Palmer, "Peru's Persistent Problems."

35. For a detailed discussion of Peru's parlous national security situation as of the early 1990s, see David Scott Palmer, "National Security."

36. The "agreement" was never consummated, because the Peruvian media denounced it immediately as giving rights to Ecuador in territory that Peru considered to be its own. In the uproar that followed, Peru's foreign minister lost his job. See St. John, "Conflict in the Cordillera del Cóndor: The Ecuador-Peru Dispute," 82–83; and Adrián Bonilla, "Ecuador-Peru: National Interests and Political Process of the 1995 Armed Conflict."

37. Palmer interview with Caryn Hollis, U.S. Defense Intelligence Agency analyst, June 23, 1992.

38. Palmer interview with Javier Ponce, deputy chief of the Ecuadorean Mission to the United Nations, February 13, 1995; Palmer interview with Edgar Terán, Ecuador's ambassador to the United States, April 3, 1996; Palmer interview with General José D. Williams Zapata (ret., the commander of the Peruvian special forces brigade in the Cenepa in 1995), December 4, 2008.

39. This was the lament of several career diplomats interviewed, both Ecuadorean and Peruvian.

40. Palmer interview with Javier Ponce. Other accounts differ. According to one, fighting started with an Ecuadorean helicopter attack on a Peruvian outpost four kilometers from the border. U.S. Embassy of Peru, *The 1995 Peruvian-Ecuadorean Border Conflict*, 3. According to Lynn Sicade, a U.S. diplomat and specialist on this case, in an interview with Palmer on April 2, 1996, no military initiative of such significance would have been carried out without explicit orders from the Peruvian high command.

41. According to one report, the conflict was costing each side $10 million per day. James Brooke, "Two Leaders Seek Laurels along Peru-Ecuador Border." Ecuador's finance minister, Modesto Correa, announced that the direct cost of the nondeclared war up to March 1 for his government was "approximately $250 million." "Finance Minister on Cost of War, Reserve Reduction." While published accounts noted between 200 and 300 total casualties on both sides, interviews with Peruvian and Ecuadorean military and diplomatic authorities put the figures much higher, between 1,000 and 1,500. Bonilla, "Ecuador-Peru: National Interests and Political Process of the 1995 Armed Conflict"; Mares, "Deterrence Bargaining in Ecuador and Peru's Enduring Rivalry: Designing Strategies around Military Weakness." Subsequent inter-

views, however, indicate that Ecuador lost fewer than 100 in the fighting, while Peru may have suffered three to four times that number. Palmer interviews with Col. Juan Villegas (Ecuador), February 13 and 28, 2008. In the 1995 war, he was in the Cenepa as a member of the personal staff of Gen. Paco Moncayo (Ecuador). These approximate figures were also confirmed in Palmer interviews with Jamil Mahuad, former president of Ecuador, July 11, 2008; and Gen. José Williams Zapata (Peru), December 4, 2008.

42. "Tiwinza: Descansa en paz"; James Brooke, "Peru and Ecuador Halt Fighting along Border, Claiming Victory." In the December 4, 2008, interview General Williams noted that Peruvian forces were successful in dislodging Ecuadorean troops from other bases and in capturing heavy artillery pieces in these operations.

43. According to Gen. Carlos Chamochumbi (ret., commander of the Peruvian army division in the border city of Tumbes in 1991–1992), this part of the border was the most difficult for the Peruvian forces to defend, given the inhospitable terrain and complete absence of roads in the area. As a result, in his view, the eventual outcome was a political but not a military solution for Peru. Palmer interview, August 14, 2008.

44. U.S. Embassy of Peru, *The 1995 Peruvian-Ecuadorean Border Conflicts*, 5.

45. Col. Glenn R. Weidner, "Peacekeeping in the Upper Cenepa Valley: A Regional Response to Crisis," 14.

46. "Government Issues Communique on Peruvian Border Incursions."

47. In the Palmer interview Col. Juan Villegas noted that President Durán Ballén's decision to return to the aegis of the Rio Protocol in the first days of the conflict was based on consultations with his ministers of foreign affairs and defense and the military joint command. At that time they did not know the state of readiness of the Peruvian forces and feared that the war could widen across the entire frontier, a situation for which their own armed forces were not prepared. See also the confirming discussion in Mares, "Deterrence Bargaining in the Ecuador-Peru Enduring Rivalry," 118.

48. "Foreign Minister Concedes Protocol with Peru Valid." President Durán Ballén's wording was ambiguous, and he never clarified it. The majority of people polled in Ecuador's two major cities believed that he did not accept the protocol. Among respondents with graduate school levels of education, the figure is even higher. Poll on January 6, 1996, in Quito and Quayaquil by *Informe Confidencial*.

49. Alfredo Luna Tobar, "Vigencia e inejecutibilidad del Protocolo de 1942," especially 9–15.

50. U.S. Embassy of Peru, *The 1995 Peruvian-Ecuadorean Border Conflict*, 4.

51. "Communique by the Guarantor Countries of the Rio Protocol of 1942 to Ecuador and Peru," Brasilia, January 27, 1995. Ambassador Alexander Watson, assistant secretary of state for American republics affairs between July 1993 and March 1996, was the U.S. representative sent to this first meeting. In an interview with Palmer (October 24, 2001), he noted that he and his guarantor colleagues immediately recognized the historic significance of the moment and all worked until 4 a.m. every day to begin to establish a framework for negotiations.

52. From material provided by ambassador Luigi Einaudi, U.S. guarantor representative, in an interview with Palmer, January 29, 1997. These key points appeared later in Luigi R. Einaudi, "The Ecuador-Peru Peace Process," 415–422.

53. *Declaración de Paz de Itamaraty entre Ecuador y Perú*, Brasilia, February 17, 1995.

54. See Weidner, "Operation Safe Border," for a detailed account of the MOMEP mission.

55. Ibid. The importance of the MOMEP mission and its members' interactions with representatives of the Ecuadorean and Peruvian military at this early juncture were critical. In spite of multiple obstacles, substantive progress was made on the military side of the situation even as the civilian side struggled with procedural issues. The MOMEP mission's early achievements provided the multilateral peacekeeping process with much of its momentum in 1995 and early 1996. Updating Weidner, but very much reinforcing these observations on the significance of the military mission for the peacekeeping process, was the presentation by Col. Leon Rios, "The Role of the Military in the Peace Process," at the conference on Security Cooperation in the Western Hemisphere: Lessons from the 1995 Ecuador-Peru Conflict, sponsored by the U.S. Southern Command, the U.S. Army War College, and the North-South Center, University of Miami, North-South Center, Miami, Florida, December 6, 1996.

56. Marcella, *War and Peace in the Amazon*, 1–2.

57. *Declaración de los países garantes del Protocolo de Rio de Janeiro de 1942 y de los Vicecancilleres del Ecuador y del Perú sobre los avances en el proceso de paz*, Brasilia, October 8, 1995.

58. *Hoy* (Quito) December 29, 1995, and February 12, 13, 14, 23, and 26, 1996; the Kfir bombers have U.S.-built engines, so their sale to third parties requires U.S. approval.

59. Gabriel Marcella, "Epilogue: The Peace of October 1998."

60. "Peru: Foreign Minister Resigns in Midst of Negotiations with Ecuador"; and a confidential Mares interview with a former high-ranking Peruvian diplomat in Lima, March 26, 1999.

Chapter 3

1. Douglass C. North, *Institutions, Institutional Change and Economic Performance.*

2. Institutional changes in either country after 1998 are not relevant to our argument about war and peace, so they are not discussed here except when they highlight a particular point of interest.

3. Massive street protests and questionable congressional resolutions have thrice toppled Ecuadorean presidents (Abdalá Bucaram in 1997, Jamil Mahuad in 2000, and Lucio Gutiérrez in 2005). Beginning in 1998, President Fujimori increasingly demanded that his majority in congress modify legislation and provide him with more leeway in policy making. He won fraudulent elections in the spring of 2000 but then was forced to flee the country in November after videotapes revealed that his top advisor, Vladimiro Montesinos, was making payoffs to members of congress.

4. Edward D. Mansfield and Jack Snyder ("Democratization and the Danger of War") develop the argument that democracies in transition from authoritarianism would be more prone to war, but their argument cannot explain how that same institutional context also produces peace.

5. Alfred Stepan, *The Military in Politics: Changing Patterns in Brazil*; Alain Rouquié, *The Military and the State in Latin America.*

6. Matthew Soberg Shugart and John M. Carey, *Presidents and Assemblies: Constitutional Design and Electoral Dynamics*, 274.

7. Catherine M. Conaghan, "Politicians against Parties: Discord and Disconnection in Ecuador's Party System"; Andrés Mejía Acosta, "¿Una democracia ingovernable?: Arreglos constitucionales, partidos políticos y elecciones en Ecuador, 1979–1996."

8. Party switching rates in the period between August 1993 and September 1993 are included because of the high percentage of party switching in that short time span.

9. Conaghan, "Politicians against Parties," 450–453.

10. Arend Lijphart, *Electoral Systems and Party Systems: A Study of Twenty-seven Democracies, 1945–1990*, 24. The d'Hondt formula uses a highest averages method for allocating seats under PR. It is named after Belgian mathematician Victor d'Hondt.

11. In a May 1990 survey 52 percent of respondents believed that the legislature had performed either "badly" or "very badly" during the García presidency of 1985–1990, and 54 percent wanted the senate and house to fuse into one chamber. APOYO S.A., Lima, May 1990, as cited in Enrique Bernales Ballesteros, *Parlamento y democracia*, 239, 260.

12. In 1998, in a controversial maneuver which entailed forcing out three electoral tribunal judges who opposed a third consecutive presidential term and allegedly making payoffs to members of congress, President Fujimori was able to get a ruling from congress that discounted his first election under the prior constitution. This unconstitutional determination allowed him to run for the presidency for a third consecutive term in 2000 by arguing that his 1995 election was his first under the new 1993 constitution.

13. Scott Mainwaring and Timothy R. Scully, "Introduction: Party Systems in Latin America," in Scott Mainwaring and Timothy R. Scully, eds., *Building Democratic Institutions: Party Systems in Latin America*, 6–7.

14. Scott Mainwaring and Matthew S. Shugart, eds., *Presidentialism and Political Parties in Latin America*.

15. Here we follow the outline of the Ecuadorean constitution provided in John M. Carey, Octavio Amorim Neto, and Matthew S. Shugart, "Appendix: Outlines of Constitutional Powers in Latin America."

16. John M. Carey and Matthew Soberg Shugart, "Calling Out the Tanks or Filling Out the Forms?" 5–11.

17. Shugart and Carey, *Presidents and Assemblies*, 149–155.

18. Mejía Acosta, "¿Una democracia ingovernable?" 227.

19. "Reforma contra Bucaram"; Francisco Rosales Ramos, "Dilema"; "Ecuador: Interim Government Calls Referendum as Supreme Court Orders Arrest of Former President."

20. "Jueces unen a políticos"; "¿Cómo despolitizar la justicia?"

21. "Congress Acts Repeatedly 'For the Last Time.'"

22. Article 202, Constitution of 1998.

23. Carol Graham, "Government and Politics," 212–217; an excellent overview is Cynthia McClintock, "Presidents, Messiahs, and Constitutional Breakdowns in Peru."

24. Articles 227–230, Constitution of 1979.

25. Articles 134–136, Constitution of 1993.

26. A majority of both houses, as per article 193, Constitution of 1979; article 108, Constitution of 1993.

27. McClintock, "Presidents, Messiahs, and Constitutional Breakdowns in Peru," 309; the number of decrees is from Samuel B. Abad Yupanqui and Carolina Garcés Peralta, "El gobierno de Fujimori: Antes y después del golpe," 103.

28. Article 118, XIX, Constitution of 1993.

29. Articles 188 and 211, Constitution of 1979; articles 104 and 137, Constitution of 1993.

30. Polls by APOYO and CPI (Compañía Peruana de Estudios de Mercado y Opinión Pública S.A.C.) as cited in Federico Prieto Celi, *El golpe*, 41; see also Carlos Iván Degregori and Carlos Rivera, "Perú 1980–1993: Fuerzas armadas, subversión y democracia."

31. Latin American constitutions provide for a variety of means by which governments can suspend constitutional guarantees in periods of domestic disorder. In addition, historically the military was often explicitly provided with the role of "guarantor" of the constitution, a task that often led civilian groups and military officers to advocate active military participation in government in times of political or economic turmoil. With the democratic transition in Latin America between 1978 and 1991, however, most of the new constitutions, including those of Ecuador and Peru, have eliminated this "guarantor" provision. See Brian Loveman, *The Constitution of Tyranny: Regimes of Exception in Latin America*.

32. There are four ideal civil-military relationships. In the first two, "civilian-dominant" and "military-dominant," one side controls the other and has the power to identify the primary threats confronting the country and the appropriate response to those threats. In the third type, "a pact among equals," the two sides share equally in determining threat and policy. In the final type, "parallel spheres of action," civilian actors and the military have separate but parallel spheres of action. See David R. Mares, "Civil-Military Relations, Democracy and the Regional Neighborhood."

33. Fernando Bustamante, "Fuerzas armadas en Ecuador: ¿Puede institutionalizarse la subordinación al poder civil?"; Anita Isaacs, *Military Rule and Transition in Ecuador, 1972–92*. This may be a typical transition for militaries seeking to leave government and focus on professional tasks. Cf. Caesar Sereseres, "The Interplay of Internal War and Democratization in Guatemala since 1982."

34. "Ecuador: Institutional Crisis Continues with Investigation of President Sixto Durán Ballén"; Adrián Bonilla, "Las imágenes nacionales y la guerra: Una lectura crítica del conflicto entre Ecuador y Perú."

35. In the controversy over Bucaram's impeachment, congress and the vice president disputed who would succeed him. Thus for a few days Ecuador had three presidents, because Bucaram rejected his ouster. The armed forces initially remained aloof despite the president's efforts to garner their political support in upholding the constitution. When they finally decided to withdraw public support for Bucaram they refused to name his successor, insisting that congress and the vice president resolve the issue. Although the armed forces, along with the U.S. Embassy, favored naming the vice president, everyone accepted the maneuvering by congress, which again circumvented the constitution and placed the legislative leader in the presidency. See *Hoy*, January 29–February 13, 1997, for discussions of the crisis, in par-

ticular "Los militares tras Rosalía" and "EE.UU. no ha intervenido en la crisis"; "Ecuador: Congress Votes to Oust President Abdalá Bucaram"; Mares interview with Adrián Bonilla, April 18, 1997.

36. John Crabtree, *Peru under García: An Opportunity Lost*, 108–112; Mares interview with Cesar Azabache (Defensoría del Pueblo), April 5, 1999.

37. Mares interview with Azabache.

38. Susan Stokes, "Peru: The Rupture of Democratic Rule," 66–71.

39. David Pion-Berlin, "From Confrontation to Cooperation: Democratic Governance and Argentine Foreign Relations"; Carlos Escudé, *Foreign Policy Theory in Menem's Argentina*.

40. Menem failed because the military juntas were accused in 1998 of kidnapping and selling the babies of political prisoners, a crime which had not been considered under the 1990 amnesty. In addition, in 2007 a federal court threw out the pardons. Patrick J. McDonnell, "Pardons Voided for Argentine 'Dirty War' Pair."

41. Teodoro Hidalgo Morey, *Las ganancias de Ecuador*.

42. Mares interview with Luis Huerta (Comisión Andina de Juristas), March 25, 1999.

43. Mejía Acosta, "¿Una democracia ingovernable?" 16–17. On neo-liberal economic reforms, see Jennifer N. Collins, "Much Ado about Nothing: The Politics of Privatization in Ecuador, 1992–1996."

44. Ibid., 18–20.

45. Ibid., 284.

46. David Scott Palmer, "The Often Surprising Outcomes of Asymmetry in International Affairs: United States–Peru Relations in the 1990s," 233.

47. Gustavo Pons Muzzo, *Estudio histórico sobre el Protocolo de Rio de Janeiro*, 364.

Chapter 4

1. David R. Mares, *Violent Peace: Militarized Interstate Bargaining in Latin America*, 3–27.

2. See Adrián Bonilla, "Las imágenes nacionales y la guerra: Una lectura crítica del conflicto entre Ecuador y Perú"; see also Carlos Espinosa, "La negociación como terapia: Memoria, identidad y honor nacional en el proceso de paz Ecuador-Perú."

3. See Charles Tilly, "War Making and State Making as Organized Crime."

4. Fernando Bustamante, "Fuerzas armadas en Ecuador: ¿Puede institucionalizarse la subordinación al poder civil?"; Adrián Bonilla, "Proceso político e intereses nacionales en el conflicto Ecuador-Perú," 37–38.

5. The strength of regional and ethnic identification and interests makes an internally generated nationalism difficult in Ecuador. A common identity based on political norms, as in the United States, is also unlikely given the weakness of Ecuador's political institutions.

6. Ronald Bruce St. John, "The Boundary between Ecuador and Peru," 16.

7. See the detailed analysis in David R. Mares, "Deterrence Bargaining in the Ecuador-Peru Enduring Rivalry: Designing Strategies around Military Weakness." Also see the comparative expenditure details in note 22 below.

8. Confidential Mares interviews with Ecuadorean military officers and with former high-ranking diplomat, August 1995; see also Mares interviews with Dr. Luis

Proaño (political advisor, Ministry of Defense, Ecuador), August 14, 1995; and Col. Luis B. Hernández (personal secretary to the minister of defense of Ecuador and commander of the Tiwintza defense during the war), August 14, 1995.

9. Gobierno de Ecuador, Ministerio de Relaciones Exteriores, *Hacia la solución del problema territorial con el Perú: Libro blanco*, 145.

10. Luis Carrera de la Torre, *El proyecto binacional Puyango Tumbes*; Carlos E. Scheggia Flores, *Origen del pueblo ecuatoriano y sus infundadas pretensiones amazónicas*, 71–73.

11. Between 1983 and 1993 Ecuador increased its number of soldiers dramatically (by half), while Peru decreased its forces (by almost one-third). Yet Peru's armed forces still outnumbered their rival by two to one. U.S. Arms Control and Disarmament Agency (ACDA), *World Military Expenditures and Arms Transfers, 1993–94*, 61, 78.

12. J. Samuel Fitch, *The Military Coup d'Etat as a Political Process: Ecuador, 1948–1966*; Anita Isaacs, *Military Rule and Transition in Ecuador, 1972–92*, 2–3.

13. Mares interview with Jaime Durán (director of *Informe Confidencial*), July 15, 1997.

14. Confidential Mares interviews with Ecuadorean military officers, August 1995; see also Mares interviews with Dr. Luis Proaño and Col. Luis B. Hernández, August 14, 1995; and Lt. Gen. Frank Vargas Pazzos, *Tiwintza: Toda la verdad*, 45–62.

15. The U.S. Arms Control and Disarmament Agency (ACDA) does not record China having delivered or made any agreements to deliver arms to Ecuador from 1979 to 1993. U.S. Arms Control and Disarmament Agency, *World Military Expenditures and Arms Transfers, 1993–94*, and *World Military Expenditures and Arms Transfers, 1985*, 133 and 140, respectively.

16. Confidential interviews by Mares with an Ecuadorean diplomat, a former high-ranking Ecuadorean diplomat, Ecuadorean military officers, a Peruvian analyst, and U.S. military analysts in August and September 1995. Ecuador's navy had been bottled up in port in 1981.

17. Stephen M. Gorman, "Geopolitics and Peruvian Foreign Policy," 80; Daniel M. Masterson, *Militarism and Politics in Latin America: Peru from Sánchez Cerro to Sendero Luminoso*, 265. During the 1995 conflict, Peru's President Fujimori maintained that Peru could escalate the conflict despite initial losses because the military government of the 1970s had stockpiled weapons in preparation for a war with Chile. "Peru Was Preparing for War with Chile, Reveals President Fujimori."

18. The poll is discussed in República del Ecuador, Ministerio de Relaciones Exteriores, *Hacia la solución*, 192–193; diplomatic sentiment is discussed in *Latin American Regional Reports*, November 14, 1991, 6; and confirmed in Mares's confidential interviews in August 1995.

19. Gustavo Pons Muzzo, *Estudio histórico sobre el Protocolo de Rio de Janeiro*, 358.

20. Jaime Durán Barga, "Actitud de los ecuatorianos frente al Perú: Estudio de opinión pública."

21. *Informe Confidencial*, January 6, 1995 (Archives).

22. ACDA's estimates of Ecuadorean military expenditures as a percent of GNP indicate that they were never large. There were slight increases under military rule (1972–1979), from 2.1 percent to 2.4 percent, except for a high of 2.9 percent in 1978. After the return of democracy in 1979, military spending levels as a percent of GDP declined to the level of the early years of military government but increased significantly after the 1981 miniwar (reaching 3.3 percent in 1983) before declining dramatically after 1987 to 1.1 percent in 1993. ACDA, *World Military Expenditures,*

1985, and *World Military Expenditures, 1993–94*, 60 and 61, respectively. If we compare Ecuador's and Peru's military spending over the decade before the 1995 war, however, a different picture emerges. Peruvian expenditures in constant 1993 dollars were 13 percent lower in 1994 than in 1985 ($730 million versus $842 million), while Ecuador's increased by 58 percent ($589 million versus $373 million). Peru's per capita military expenditures between these same years declined by 31 percent (from $45 to $31), while Ecuador's increased by 33 percent (from $40 to $53). As a percent of GDP in 1985 and 1994, Peru's military expenses declined by 60 percent (from 4.5 percent to 1.8 percent), while Ecuador's increased by 78 percent (1.8 percent to 3.2 percent). Institute for Strategic Studies, *Military Balance, 1995–96*.

23. Mares, "Deterrence Bargaining."

24. Fernando Tuesta Soldevilla, *Perú político en cifras*; Raúl P. Saba, *Political Development and Democracy in Peru: Continuity in Change and Crisis*, 72–76; Philip Mauceri, *State under Siege: Development and Policy Making in Peru*, 46–58; Julio Cotler, "Political Parties and the Problems of Democratic Consolidation in Peru," 337–343.

25. When the military command requested President Belaúnde's authorization to launch a general attack on Ecuador (including a strike on the Ecuadorean fleet), however, he refused. Palmer interview with Gen. Carlos Chamochumbi (Peru, ret.), August 14, 2008.

26. Saba, *Political Development and Democracy in Peru* 137. The other authors cited in note 24 do not even mention the 1981 miniwar in their discussions of political protests against Belaúnde. Edward Schumacher, "Behind Ecuador War, Long-Smoldering Resentment."

27. Soldevilla, *Perú político en cifras*, 68.

28. Mauceri, *State under Siege*, 59–77; John Crabtree, *Peru under García: An Opportunity Lost*, 69–93, 121–183.

29. Soldevilla, *Perú político en cifras*, 64, for 1990; and David Scott Palmer, "Peru's 1995 Elections," for 1995.

30. Soldevilla, *Perú político en cifras*, 23, Table 2, and 149.

31. In March 1999 (nine years after assuming office, including two years of authoritarian government) Fujimori's approval ratings were still 64.5 percent. "En popularidad, Fujimori cede terreno a Andrade," A5.

32. Polling data from APOYO S.A. (Lima), #62001, January 1994, questions 9.1 and 9.2; and #62004, April 1994, questions 6.1 and 6.2.

33. *Hoy* (Guayaquil), March 22, 1995, and April 23, 1995; Abraham Lamas, "Ecuador-Perú: Quién ganó y quién perdió en la Guerra Empatada."

34. APOYO S.A. (Lima), "Elections," #63004, February 1995, questions 2, 3, 4, 6, and 7; "Political and Economic Situation," #62002, February 1995, questions 2.1, 2.2, and 2.5; and "Political and Economic Situation," #62003, March 1995, questions 4.1a, 4.1c, 4.1e, and 4.1f.

35. "Ecuador: Institutional Crisis Continues with Investigation of President Sixto Durán Ballén"; Bonilla, "Las imágenes nacionales y la guerra."

36. República del Ecuador, Ministerio de Relaciones Exteriores de Ecuador, *Misión en Washington*, 61.

37. Mares confidential interview with an Ecuadorean diplomat and with a former high-ranking Ecuadorean diplomat, August 1995.

38. Mares confidential interview with a former high-ranking Ecuadorean diplomat, August 1995. Mares's interviews with Gen. Frank Vargas Pazzos, August 16,

1995, and Gen. José W. Gallardo Román (minister of defense), August 1995, indicate that some military officers have reinterpreted the events of 1981 to suggest that they were holding their own and that civilian authorities capitulated. Army chief general Paco Moncayo, however, claimed that the armed forces were not ready to defend themselves against a broad Peruvian attack in 1981 as a result of having focused on governing the country between 1973 and 1979 (Mares interview August 16, 1995).

39. Confidential Mares interview with an Ecuadorean diplomat, August 1995. A more specific perspective on the decision is provided by Col. Juan Villegas of the Ecuadorean army, who was a member of army chief general Paco Moncayo's personal staff as a major during the first months of the 1995 war. He noted that President Durán Ballén's decision to return to the aegis of the Rio Protocol as the conflict was breaking out was the product of prior consultation with key officials in the Ministry of Foreign Relations, Ministry of Defense, and Joint Command and was based on a lack of certainty about the state of readiness of the Peruvian armed forces and fear that they might expand the war across the entire frontier (Palmer interview, February 13, 2008).

40. See the discussion in *NotiSur* during those years.

41. Mares interview with Dr. Luis Proaño, August 14, 1995.

42. "Government Issues Communique on Peruvian Border Incursions."

43. Mares interview with Jaime Durán, July 15, 1997.

44. Pons Muzzo, *Estudio histórico sobre el Protocolo de Rio de Janeiro*, 364.

45. "Guerra avisada no mata gente, las advertencias del Gral. Salinas Sedó"; "El caballero de las dos torres"; see also Francisco Carrión Mena, *La paz por dentro, Ecuador-Perú: Testimonio de una negociación*, 44–45.

46. *Hoy*, March 22 and April 3, 1995; Lamas, "Ecuador-Perú."

47. *NotiSur—Latin American Political Affairs*, March 10, 1995.

Chapter 5

1. Midway through 1997 foreign investment in Ecuador had fallen 59 percent over the previous year, Gross Domestic Product (GDP) growth was projected at only 1.0 percent for the year and inflation at 31 percent, giving Ecuador the second highest inflation rate in the region. "Economic Policies Criticized at Home and Abroad."

2. John M. Owen, "How Liberalism Produces Democratic Peace"; William J. Dixon, "Democracy and the Peaceful Settlement of International Conflict"; John R. Oneal and Bruce M. Russett, "The Classical Liberals Were Right: Democracy, Interdependence, and Conflict, 1950–1985."

3. Pedro Saad Herrería, *La caída de Abdalá*, 137–138.

4. APOYO S.A. (Lima), "Elections," #63004, February 1995, question 5; "Political and Economic Situation," #62002, February 1995, question 2.4; and "Political and Economic Situation," #62003, March 1995, questions 4.2 and 4.3.

5. Charles F. Hermann, Charles W. Kegley, and James N. Rosenau, eds., *New Directions in the Study of Foreign Policy*; Jonathan Bendor and Thomas H. Hammond, "Rethinking Allison's Models."

6. Fernando Bustamante, "Ecuador: Putting an End to Ghosts of the Past?" 208–212.

7. Francisco Carrión Mena, *La paz por dentro, Ecuador-Perú: Testimonio de una negociación*, 238–265.

8. Ibid., 372–378.

9. Ronald Bruce St. John, *The Foreign Policy of Peru*, 3–4; Julio Alvarez Sabogal, *La política exterior del Fujimorato (1990–2000)*, 102–103.

10. At the time, allegations of inappropriate personal behavior (corruption and homosexuality) were among the causes that Fujimori cited for the abrupt and large-scale dismissal. Only after his departure from the presidency in 2000 were the individuals affected exonerated, compensated, and restored to diplomatic service if still eligible. Of the eight foreign ministers during the Fujimori years, only the first, ambassador Luis Marchand Stens, was a career diplomat. Alvarez Sabogal, *La política exterior del Fujimorato*, 103–109.

11. On the multiple domestic crises, see David Scott Palmer, "Peru's Persistent Problems." On the diplomatic initiatives, see Hernando Burgos, "Conflicto Perú-Ecuador: El gran viraje," 17.

12. David Scott Palmer, "Peru-Ecuador Border Conflict: Missed Opportunities, Misplaced Nationalism, and Multilateral Peacekeeping," 115 and n. 10. Also Burgos, "Conflicto Perú-Ecuador," 18–19.

13. Catherine M. Conaghan, *Fujimori's Peru: Deception in the Public Sphere*, 93–94, 140–162.

14. *Latin American Weekly Report* WR-98-39 (October 8, 1998), 461, and WR-96-43 (November 3, 1998), 509.

15. Fitch's detailed analysis of military perceptions and justifications for supporting or threatening Ecuadorean democratic governments in the period 1948 to 1966 does not uncover disagreements between civilians and military officers over the Amazonian issue. J. Samuel Fitch, *The Military Coup d'Etat as a Political Process: Ecuador, 1948–1966*; Anita Isaacs, *Military Rule and Transition in Ecuador, 1972–92*, 2–3, also does not reference the question of the border in her summary of why the military grew disenchanted with democracy in the 1960s.

16. Isaacs, *Military Rule and Transition in Ecuador, 1972–92*, 109–111.

17. See note 22 of Chapter 4 for sources and details on Ecuador's military expenditures as a percentage of GNP, which declined by over half from the 1970s under military rule and the early 1990s under civilian government. The increase of institutional capacity is noted in the contracting of Israeli and Chilean intelligence and communications exports and the significantly more effective planning that went into the incursion into disputed territory in 1995 than in 1991, including getting the navy dispersed from port before the actual fighting began. Confidential Mares interviews with an Ecuadorean diplomat, a former high-ranking Ecuadorean diplomat, Ecuadorean military officers, a Peruvian analyst, and U.S. military analysts in August and September 1995. Army Chief General Moncayo claimed that the armed forces were not ready to defend the country against a broad Peruvian attack in 1981, as a result of having focused on governing the country between 1973 and 1979 (Mares interview, August 16, 1995). Ecuador's military expenditures increase in comparison to Peru's between 1985 and 1994 (see note 22 of Chapter 4 for details). This only reinforces General Moncayo's claim.

18. For a detailed analysis, see David R. Mares, "Deterrence Bargaining in the Ecuador-Peru Enduring Rivalry: Designing Strategies around Military Weakness."

19. This analysis is based on responses by army and air force officers to the Mares presentation "La disuasión y el conflicto Ecuador-Perú"; interviews with Chief General Moncayo and Adrián Bonilla; and public opinion data analyzed in this chapter and previous chapters.

20. This is our conclusion, drawn from comments that surfaced repeatedly in various interviews by Mares with Ecuadorean military officers during August 1995.

21. Mares interview with Gen. Frank Vargas Pazzos, August 16, 1996.

22. Isaacs, *Military Rule and Transition in Ecuador, 1972–92*, 131–138.

23. Víctor Villanueva, *El CAEM y la revolución de la fuerza armada*.

24. See, among others, David Scott Palmer, *Revolution from Above: Military Government and Popular Participation in Peru, 1968–1972*, 42–77.

25. Daniel M. Masterson, *Militarism and Politics in Latin America: Peru from Sánchez Cerro to Sendero Luminoso*, 203–235.

26. Philip Mauceri, *State under Siege: Development and Policy Making in Peru*, 15–36.

27. Cynthia McClintock, "Peru: Precarious Regimes, Authoritarian and Democratic," 349–351.

28. Palmer, "Peru's Persistent Problems."

29. Enrique Obando, "Fujimori y las fuerzas armadas," 363–367; also Palmer interview with Gen. Carlos Chamochumbi (ret.), August 14, 2008.

30. David Scott Palmer, "Revolution in the Name of Mao: Rebellion and Response in Peru."

31. Douglass C. North, *Institutions, Institutional Change and Economic Performance*; Sven Steinmo, Kathleen Thelen, and Frank Longstreth, eds., *Structuring Politics: Historical Institutionalism in Comparative Analysis*.

32. Brian Barry, *Sociologists, Economists & Democracy*, 99–100.

33. "Ecuador: President Sixto Durán Ballén Suffers Defeat in Referendum on Constitutional Reforms."

34. Mares, "Deterrence Bargaining in the Ecuador-Peru Enduring Rivalry."

35. "Ecuador: Sixto Durán Ballén Turns Over Presidency to Successor Abdalá Bucaram."

36. "Opinión: Buenas intenciones." Cf. "¿De qué pedimos perdón?" Bucaram attempted to link his overthrow to his visit to Peru. "Caída no tuvo vínculo con viaje"; "Opinión: Peregrinaje irresponsable"; and "Bucaram merece castigo, dice Gustavo Noboa."

37. "Opinión: Ministerio de Etnias," October 25, 1996.

38. "Cuidado con el Art. 100." Once Bucaram was discredited and booted out of office, congress and the courts found him mentally competent enough to stand trial for corruption as well as potentially popular enough to win another election, so they stripped him of his constitutional right to run for public office. Congress also expelled thirteen of his supporters on corruption charges; it appears that no member of congress who opposed Bucaram was so treated. "No caminan las reformas"; "Habrá nuevas descalificaciones."

39. Kenneth P. Jameson, "Dollarization in Latin America: Wave of the Future or Flight to the Past?"

40. Robert Andolina, "The Sovereign and Its Shadow: Constituent Assembly and Indigenous Movement in Ecuador."

41. David Scott Palmer, "Peru's 1995 Elections."

42. Carol Wise, "Against the Odds: The Paradoxes of Peru's Economic Recovery in the 1990s," 209–217 and Table 3, 212.

43. Among others, see Palmer, "Revolution in the Name of Mao," 204–216.

44. While non-Ecuadorean NGOs can help their Ecuadorean counterparts, the effort cannot be perceived as an international one. On the dangers of domestic NGOs being perceived as local branches of international forces, see Sergio Aguayo Quezada, "Del anonimato al protagonismo."

Chapter 6

1. William L. Krieg, *Ecuadorean-Peruvian Rivalry in the Upper Amazon*, Enlarged to Include the Paquisha Incident (1981), 16–115.

2. The treaty's official English translation may be found in ibid., Appendix I.

3. David Scott Palmer, "El conflicto Ecuador-Perú: El papel de los garantes," 48–55.

4. Krieg, *Ecuadorean-Peruvian Rivalry in the Upper Amazon*.

5. A personal experience provides a telling anecdote with regard to Ecuador's nationalistic sentiment during these years. When Palmer went to mail a letter abroad from a Quito post office in 1964, he was told it would not be sent unless he affixed a stamp with a map of Ecuador that included the Amazon and the words "El Amazón es nuestro" (The Amazon is ours).

6. For MOMEP mission details, see Col. Glenn R. Weidner, "Peacekeeping in the Upper Cenepa Valley: A Regional Response to Crisis," 45–64.

7. As reported by *NotiSur* from Mexico City, February 2, 1996.

8. "Substantive impasses" (*impases subsistentes*) was a linguistic innovation by the guarantors in the Declaration of Itamaraty, which permitted both Peru and Ecuador to get beyond the legalisms of their individual interpretations. We are indebted to a key Peruvian participant in the early diplomatic discussions who prefers not to be named for this observation, as expressed in an August 22, 1996, letter to Palmer.

9. The explanatory information in brackets was most helpfully provided to Palmer in 1997 by two U.S. government analysts who have followed the conflict closely but prefer not to be identified.

10. República del Ecuador, Ministerio de Relaciones Exteriores, *Ecuador: Impases subsistentes*; public release, March 6, 1996.

11. República del Perú, Ministerio de Relaciones Exteriores, *Ayuda memoria: Desacuerdos sobre demarcación de la frontera*; public release, March 6, 1996.

12. As one account has it, Peruvian foreign minister Francisco Tudela gave a bombastic and insulting presentation at the Santiago meeting. Forewarned, his Ecuadorean counterpart, Galo Leoro, responded in a measured but barely controlled tone. The guarantors halted the exchange and sent the parties to their separate rooms. Rather than risk another outburst, the guarantors then engaged in lengthy "shuttle diplomacy" between the two parties' quarters with a draft document that eventually became the breakthrough Santiago Agreement. Palmer interview with one of the meeting participants, a Latin American diplomat who requested not to be named, August 23, 1997.

13. *Acuerdo de Santiago* (Chile), October 29, 1996.

14. Luigi R. Einaudi, "The Ecuador-Peru Peace Process," 411–415.

15. *Comunicado de Prensa* (Buenos Aires, Argentina), June 19, 1996.

16. From one Peruvian career diplomat's viewpoint, the main obstacle to definitive resolution of the boundary problem was its politicization over time, with a corresponding lack of emphasis on its juridical parameters—specifically the formal

boundary treaty (Rio Protocol) and an arbitration decision (the Braz Dias de Aguiar Award), both of which Peru and Ecuador accepted and almost completely executed. The definitive designation of the border in the areas where it was not yet demarcated, from this perspective, required nothing less than the application of the international legal instruments establishing it in the first place. In this view, therefore, Ecuador's position constituted a continuing violation of international agreements in effect and was the principal source of the recurring bilateral conflicts. Letter to Palmer, August 26, 1996.

17. Summarized in Carlos Espinosa, "La negociación como terapia: Memoria, identidad y honor nacional en el proceso de paz Ecuador-Perú," 128.

18. Although the Rio Protocol contained no explicit arrangement for guarantor arbitration, the October 1996 Santiago Accord provision stipulating that the guarantors would "enforce agreements and propose solutions when the parties cannot agree among themselves" (as noted in the text above) served as the basis for the agreement among the parties to apply the provision at this critical juncture.

19. This is an encapsulated summary of the key provisions of the letter, dated October 23, 1998, sent by the four presidents of the guarantor countries to presidents Menem and Fujimori, to which the latter had previously agreed. The complete text is found in Francisco Carrión Mena, *La paz por dentro, Ecuador-Perú: Testimonio de una negociación*, 559–562.

20. See the general discussion of the array of multilateral initiatives conducive to conflict resolution in the hemisphere in Francisco Rojas Aravena, "América Latina: Alternativas y mecanismos de prevención en situaciones vinculadas a la soberanía territorial."

21. In many ways Ambassador Einaudi was the perfect person to serve as the U.S. representative with the guarantors. He had worked on Latin American civil-military affairs at the Rand Corporation in the 1960s and early 1970s after completing a dissertation on the Peruvian military at Harvard. Invited by Henry Kissinger to join the State Department's Office of Policy Planning in 1973 as a political appointee, he continued through both Republican and Democratic administrations as the director of the Policy Planning Office of the Bureau of Inter-American Affairs and joined the Senior Executive Service (SES) in the early 1980s. Trusted for his knowledge of the issues and his balanced judgment, he became the institutional memory on Latin American matters within the bureau. In 1989 President George Herbert Walker Bush appointed him ambassador to the OAS, where he served with distinction until 1993, when he returned to the secretary of state's Office of Policy Planning. Although Einaudi officially retired from the State Department in 1997, he continued to serve as the U.S. representative with the guarantors. He was elected by OAS members in 2000 as assistant secretary general of that organization and served in that position until 2003.

22. Former president of Ecuador Jamil Mahuad, however, indicated that he was the one who originally made this novel proposal as a way to make a final agreement more acceptable to key actors and the public in Ecuador (Palmer interview, July 11, 2008). But the detailed analysis of the process by a leading Ecuadorean participant makes it clear that the only palatable mechanism by which such a solution could be presented was through the agreement by both presidents that the guarantors be the ones who proposed it and that their proposal be binding on both parties. Carrión Mena, *La paz por dentro, Ecuador-Perú*, 488–567. Given Ambassador Einaudi's cen-

tral role in the process, President Mahuad's statement is not at odds with the view that the U.S. guarantor representative drafted the symbolic territorial proposal into the formal document.

23. These tributes to Ambassador Einaudi were among those made by presidents Alberto Fujimori of Peru and Jamil Mahuad of Ecuador at a ceremony attended by Palmer in Washington, D.C., in February 1999 to bestow on Dr. Einaudi their countries' highest medals to foreign citizens.

24. Krieg, *Ecuadorean-Peruvian Rivalry in the Upper Amazon*, 26–76.

25. Ibid., 143–144, 151.

26. Ecuador viewed these elements as possibly advantageous to its own position. Argentina, traditionally an ally of Peru, might be neutralized in its guarantor role by this controversy. Chile, given its historic tensions with Peru, might favor Ecuador's position as a guarantor. In fact, these eventualities did not come into play due to the willingness of the guarantor representatives to allow U.S. representative Luigi Einaudi to play a leading role and to leave Brazil and President Fernando Henrique Cardoso as guarantor coordinator. See the discussion below.

27. These observations are derived from various Palmer interviews between 1995 and 1998 with several of the participants in the process and analysts who prefer not to be named.

28. Palmer confirmed Dr. Ferrero Costa's hard-line position in an interview with him on September 9, 1997.

29. Interview with Fernando Aguayo on *Primera Hora*.

30. President Jamil Mahuad was especially effusive in his praise of the behind-the-scenes role of Brazil's President Cardoso in the interview with Palmer, July 11, 2008. Details of these critical moments are presented in Francisco Carrión Mena, *La paz por dentro, Ecuador-Perú*, 538–545.

31. David Scott Palmer, "El conflicto Ecuador-Perú: El papel de los garantes," 46–47.

32. Espinosa, "La negociación como terapia," 114–128.

33. A Cedatos-Gallup poll taken in major Ecuadorean cities on October 17, 1998, just after the Ecuadorean congress approved the terms of the settlement, showed a 71 percent level of support, with just 20 percent opposed. Carrión Mena, *La paz por dentro*, 535.

34. Carlos Espinosa, "Memory, Identity, and Negotiation Dynamics."

Chapter 7

1. Even Beth A. Simmons, "Trade and Territorial Conflict in Latin America," whose analysis claims that the economic benefits of peace are such that the countries should have resolved their dispute, could only come up with a figure of an average loss in bilateral trade of $35 million per year in the absence of a settlement. In 1995 Ecuador's GDP was $18 billion while Peru's was $59 billion. International Monetary Fund (IMF), *World Economic Outlook Database*.

2. As the street demonstrations in Argentina, Bolivia, and Ecuador that overthrew democratically elected presidents who adopted unpopular policies make clear, policy is not just a matter of political will.

3. Comunidad Andina, Secretaría General, Sistema Integrado de Comercio Exterior, Proyecto—Estadística, *El comercio exterior de la Comunidad Andina, 1969–2007*

(2008), 10, http://www.comunidadandina.org/documentos.asp (accessed October 10, 2008).

4. Ibid., 12.

5. Oscar Maúrtua de Romaña, "Hace cinco años estalló la paz," 76.

6. Peru Chapter, *Binational Development Plan for the Peru-Ecuador Border Region, 2000–2006* (Lima: Comisión Binacional, 2007), 10.

7. For Peru, $277 million of the total came from the Peruvian government, $115 million from foreign governments, $47 million from NGOs, and $846 million from the private sector in concessions, mostly for road construction. Peru Chapter, *Binational Development Plan*, 7, 15, 22–23. For Ecuador, about 20 percent of total expenditures were provided by international contributions, with the rest coming from the national government. República del Ecuador, "Plan Binacional Capítulo Ecuador," 1.

8. Palmer interview with Edith Alcorta Silva Santisteban and Franklin Rojas Escalante of the Peruvian office of the Binational Development Plan, August 15, 2008; and Peru Chapter, *Binational Development Plan*.

9. República del Ecuador, "Plan Binacional Capítulo Ecuador," Programa B, 1–2.

10. *El Comercio* (Lima), May 8, 2008.

11. *El Comercio* (Lima), September 12, 2008.

12. Palmer interviews with Alcorta and Rojas, August 15, 2008.

13. *Peru 21* (Lima), January 18, 2008. Former foreign minister of Peru Fernando Trazegnies, however, has stated that the sea boundary issue is with Chile alone and that the possibility of a problem with Ecuador is "an invention of the Peruvian press" (statement at the Seminario Internacional Ecuador-Perú: Una Década de Paz, FLACSO Ecuador, Quito, October 28, 2008).

14. International Court of Justice, "Case concerning the Dispute Regarding Navigational and Related Rights (Costa Rica v. Nicaragua)," judgment, July 13, 2009; Ana Núñez, "Ecuador firma tratado de límites marítimos con nuestro país."

15. "Comandante de la Armada de Chile inicia hoy inédita visita a Bolivia"; Alexia Vlahos, "Bolivia Demands Access to Pacific Ocean: Arica Tunnel."

16. Matt Glenn, "ICJ Begins Hearings in Argentina-Uruguay Paper Mill Dispute."

17. "Pulp-Mill Verdict Satisfies Honour on Both Sides of River Uruguay," 1.

18. José Rodríguez Elizondo, *Chile-Perú: El siglo que vivimos en peligro*.

19. The use of force is commonly ranged along a five-point scale: no use of force, a verbal threat, display of force, its use resulting in fewer than 1,000 battlefield-related deaths, and its use causing more deaths (the "technical" definition of war among quantitatively oriented social scientists). Included in the use of force is the seizure of one of the disputants' assets (people, fishing vessels, oil platforms, and so forth) by the authorized agents (customs officials, police, coast guard, national guard, military) of another disputant if the latter responded with some type of protest (diplomatic, economic, or military). The category display of force includes over-flights by military aircraft or hot pursuits across a border (for example, Colombian forces pursuing Colombian guerrillas) that produce a protest by the country whose territory is penetrated. Thus the action is not classified as militarized if the official agents of one country inadvertently cross the border and are escorted back without shots being fired, arrests made, or official protests by the country whose territory was penetrated (as has been occurring more frequently on the U.S.-Mexico border). Use of force by nonstate actors is not included (for example, Colombian paramilitaries attacking Panamanian villages).

20. Raúl Zibechi, "Is Brazil Creating Its Own 'Backyard'?" After multiple exchanges and conversations between President Lugo and President da Silva, Brazil agreed in late 2009 to pay a higher price to Paraguay for the Itaipú-generated electricity.

21. "Nicaraguan Missiles and Honduran Warplanes on Political Collision Course."

22. Council on Hemispheric Affairs, "Memorandum to the Press: Venezuela's Security Factors and Foreign Policy Goals."

23. Susan Abad, "Fábricas de armas y municiones de Venezuela desequilibrarían la región."

24. Simon Romero, "Venezuela Spending on Arms Soars to World's Top Ranks."

25. Jeremy McDermott, "U.S. Condemns Bolivia and Venezuela Ambassador Expulsions as 'Grave Error.'"

26. John Lindsay-Poland, "New Military Base in Colombia Would Spread Pentagon Reach throughout Latin America"; Simon Romero, "Increased U.S. Military Presence in Colombia Could Pose Problems with Neighbors."

27. Angus Reid, "Colombia, Ecuador Assess Cross-Border Incursion."

28. BBC News, "Troop Pull-out Urged in Nicaragua–Costa Rica Border Row"; Nicky Pear and Alexandra Reed, "Dredging Up an Old Issue: An Analysis of the Longstanding Dispute between Costa Rica and Nicaragua over the San Juan River"; Alex Sánchez, "Costa Rica: An Army-less Nation in a Problem-Prone Region."

29. See the analysis in David R. Mares, "Deterrence Bargaining in the Ecuador-Peru Enduring Rivalry: Designing Strategies around Military Weakness."

30. It had long been assumed that foreign policy had little impact on U.S. presidential elections, but recent work is calling this into question. John H. Aldrich, John L. Sullivan, and Eugene Brogida, "Foreign Affairs and Issue Voting: Do Presidential Candidates 'Waltz before a Blind Audience'?" A historical case is made in Charles P. Korr, *Cromwell and the New Model Foreign Policy: England's Policy toward France, 1649–1658.*

Appendix A

The collaboration of Octavio Amorim Neto on this appendix is gratefully acknowledged.

1. Markku Laakso and Rein Taagepera, "'Effective' Number of Parties: A Measure with Application to West Europe"; Gary Cox, *Making Votes Count: Strategic Coordination in the World's Electoral Systems*; Arend Lijphart, *Electoral Systems and Party Systems: A Study of Twenty-seven Democracies, 1945–1990*; Rein Taagepera and Matthew Sobert Shugart, *Seats and Votes: The Effects and Determinants of Electoral Systems.*

2. Rein Taagepera, "Expansion and Contraction Patterns of Large Polities: Context for Russia," *International Studies Quarterly* 41 (1997), 475–504 (formula on 478).

BIBLIOGRAPHY

Abad, Susan. "Fábricas de armas y municiones de Venezuela desequilibrarían la región." *El Comercio* (Peru, December 12, 2006). http://elcomercio.pe/edicionimpresa/html/2006-12-17/ImEcMundo0634393.html (accessed July 6, 2011).

Abad Yupanqui, Samuel B., and Carolina Garcés Peralta. "El gobierno de Fujimori: Antes y después del golpe." In Comisión Andina de Juristas, ed., *Del golpe de estado a la nueva constitución*, 85–190. Series: Lecturas sobre temas constitucionales 9. Lima: Comisión Andina de Juristas, 1993.

"Acuerdo de Santiago" (Santiago, Chile, October 29, 1996).

Aguayo, Fernando. Interview on *Primera Hora*. Channel 3 (Quito, June 16, 1998; transcript copy given to Mares).

Aguayo Quezada, Sergio. "Del anonimato al protagonismo." *Foro Internacional* 127 (January–March 1992), 324–341.

Aldrich, John H., John L. Sullivan, and Eugene Brogida. "Foreign Affairs and Issue Voting: Do Presidential Candidates 'Waltz before a Blind Audience'?" *American Political Science Review* 83, no. 1 (March 1989), 123–142.

Alexandroff, Alan, and Richard Rosecrance. "Deterrence in 1939." *World Politics* (April 1977), 404–424.

Alvarez Sabogal, Julio. *La política exterior del Fujimorato (1990–2000)*. Lima: Editor-Autor Julio Alvarez, 2009.

Anderson, Thomas P. *The War of the Dispossessed: Honduras and El Salvador, 1969*. Lincoln: University of Nebraska Press, 1981.

Andolina, Robert. "The Sovereign and Its Shadow: Constituent Assembly and Indigenous Movement in Ecuador." *Journal of Latin American Studies* 35, no. 4 (November 2003), 721–750.

APOYO S.A. "Adults of Metropolitan Lima," #62001 (Lima, January 1994), via Roper Center for Public Opinion Research, University of Connecticut, Storrs.

———. "Adults of Metropolitan Lima," #62004 (Lima, April 8–11, 1994), via Roper Center for Public Opinion Research, University of Connecticut, Storrs.

———. "Elections," #63004 (Lima, February 1995), via Roper Center for Public Opinion Research, University of Connecticut, Storrs.

———. "Political and Economic Situation," #62002 (Lima, February 1995), via Roper Center for Public Opinion Research, University of Connecticut, Storrs.

———. "Political and Economic Situation," #62003 (Lima, March 1995), via Roper Center for Public Opinion Research, University of Connecticut, Storrs.

Area, Leandro, and Elke Nieschulz de Stockhausen. *El Golfo de Venezuela: Documentación y cronología, Vol. II (1981–1989)* Caracas: Universidad Central de Venezuela, 1991.

Avery, William P. "Origins and Consequences of the Border Dispute between Ecuador and Peru." *Journal of Inter-American Economic Affairs* 38, no. 1 (Summer 1984), 65–77.

Baldwin, David A. *Economic Statecraft*. Princeton: Princeton University Press, 1985.

Barry, Brian. *Sociologists, Economists & Democracy*. Chicago: University of Chicago Press, 1978.

BBC News. "Troop Pull-out Urged in Nicaragua–Costa Rica Border Row," November 14, 2010. http://www.bbc.co.uk/news/world-latin-america-11751727 (accessed June 26, 2011).

Bendor, Jonathan, and Thomas H. Hammond. "Rethinking Allison's Models." *American Political Science Review* 86, no. 2 (June 1992), 301–322.

Bernales Ballesteros, Enrique. *Parlamento y democracia*. Lima: Constitución y Sociedad, 1990.

Bertram, Geoffrey. "Peru 1930–60." In Leslie Bethel, ed., *The Cambridge History of Latin America*, 8:385–449. 9 vols. Cambridge: Cambridge University Press, 1991.

"Binational Development Plan for the Peru-Ecuador Border Region, 2000–2006." Comisión Binacional, Lima, 2007.

Bonilla, Adrián. "Ecuador-Peru: National Interests and Political Process of the 1995 Armed Conflict." Paper presented to the 19th International Congress of the Latin American Studies Association, Washington, DC, September 28–30, 1995.

———. "Las imágenes nacionales y la guerra: Una lectura crítica del conflicto entre Ecuador y Perú." Paper presented at the 21st Annual Meeting of the Latin American Studies Association, Guadalajara, Mexico, April 17–19, 1997.

———. "Proceso político e intereses nacionales en el conflicto Ecuador-Perú." *Nueva Sociedad* (Caracas) 143 (May–June 1996), 30–40.

Brooke, James. "Peru and Ecuador Halt Fighting along Border, Claiming Victory." *New York Times* (February 15, 1995).

———. "Two Leaders Seek Laurels along Peru-Ecuador Border." *New York Times* (February 9, 1995).

"Bucaram merece castigo, dice Gustavo Noboa." *Hoy*, February 20, 1997.

Bunce, Valerie. *Do New Leaders Make a Difference?* Princeton: Princeton University Press, 1981.

Burgos, Hernando. "Conflicto Perú-Ecuador: El gran viraje." *Quehacer* (Lima, Centro de Estudios y Promoción del Desarrollo) 93 (January–February 1995), 6–26.

Burr, Robert N. *By Reason or Force* (1965). Berkeley: University of California Press, 1965, 1974.

Bustamante, Fernando. "Ecuador: Putting an End to Ghosts of the Past?" *Journal of Interamerican Studies and World Affairs* 34, no. 4 (Winter 1992–1993), 205–213.

―――. "Fuerzas armadas en Ecuador: ¿Puede institucionalizarse la subordinación al poder civil?" In Alexandra Vela Puga, ed., *Democracia y fuerzas armadas en Sudamerica*, 129–160. Quito: Corporación de Estudios para el Desarrollo/CORDES, 1988.

"Caída no tuvo vínculo con viaje." *Hoy*, February 19, 1997.

Carey, John M., Octavio Amorim Neto, and Matthew S. Shugart. "Appendix: Outlines of Constitutional Powers in Latin America." In Scott Mainwaring and Matthew S. Shugart, eds., *Presidentialism and Democracy in Latin America*, 440–460. Cambridge: Cambridge University Press, 1997.

Carey, John M., and Matthew Soberg Shugart. "Calling Out the Tanks or Filling Out the Forms?" In John M. Carey and Matthew Soberg Shugart, eds., *Executive Decree Authority*, 1–29. Cambridge: Cambridge University Press, 1998.

Carrera de la Torre, Luis. *El proyecto binacional Puyango Tumbes.* Quito: Asociación de Funcionarios y Empleados del Servicio Exterior [AFESE], 1990.

Carrión Mena, Francisco. *La paz por dentro, Ecuador-Perú: Testimonio de una negociación.* Quito: Imprenta Mariscal, 2008.

Centeno, Miguel Angel. *Blood and Debt: War and the Nation State in Latin America.* University Park: Pennsylvania State University Press, 2002.

Central Intelligence Agency (CIA). *The World Factbook 2010.* http://www.cia.gov/library/publications/the-world-factbook/index.html (accessed May 6, 2010).

Collins, Jennifer N. "Much Ado about Nothing: The Politics of Privatization in Ecuador, 1992–1996." Paper presented at the Annual Meeting of the Western Political Science Association, Tucson, AZ, March 13–15, 1997.

"Comandante de la Armada de Chile inicia hoy inédita visita a Bolivia." *El Mercurio* (Santiago), May 7, 2007, Observatorio Cono Sur de Defensa y Fuerzas Armadas: Informe Chile No. 249. http://www.cee-chile.org/resumen/chile/chi251-300/semchi286.htm (accessed May 22, 2007).

"¿Cómo despolitizar la justicia?" *Hoy*, April 18, 1997.

Conaghan, Catherine M. *Fujimori's Peru: Deception in the Public Sphere.* Pittsburgh: University of Pittsburgh Press, 2005.

―――. "Politicians against Parties: Discord and Disconnection in Ecuador's Party System." In Scott Mainwaring and Timothy R. Scully, eds., *Building Democratic Institutions: Party Systems in Latin America*, 434–458. Stanford: Stanford University Press, 1995.

"Congress Acts Repeatedly 'For the Last Time.'" *NotiSur* (Latin American Data Base) 7, no. 32 (September 5, 1997).

Cotler, Julio. "Political Parties and the Problems of Democratic Consolidation in

Peru." In Scott Mainwaring and Timothy Scully, eds., *Building Democratic Institutions: Party Systems in Latin America*, 323–353. Stanford: Stanford University Press, 1995.

Council on Hemispheric Affairs. "Memorandum to the Press: Venezuela's Security Factors and Foreign Policy Goals" (May 2, 2007). http://www.coha .org/2007/05/02/memorandum-to-the-press-venezuelas-security-factors-and-policy-goals (accessed May 21, 2007).

Cox, Gary. *Making Votes Count: Strategic Coordination in the World's Electoral Systems*. Cambridge: Cambridge University Press, 1997.

Crabtree, John. *Peru under García: An Opportunity Lost*. Pittsburgh: University of Pittsburgh Press, 1992.

"Cuidado con el Art. 100." *Hoy*, February 24, 1997.

Deborah, Avant D. *Political Institutions and Military Change: Lessons from Peripheral Wars*. Ithaca: Cornell University Press, 1994.

"Declaración de los países garantes del Protocolo de Rio de Janeiro de 1942 y de los Vicecancilleres del Ecuador y del Perú sobre los avances en el proceso de paz" (Brasilia, October 8, 1995).

"Declaración de Paz de Itamaraty entre Ecuador y Perú" (Brasilia: Guarantor Countries, Peru, and Ecuador, February 17, 1995).

Degregori, Carlos Iván, and Carlos Rivera. "Perú 1980–1993: Fuerzas armadas, subversión y democracia." In *Documento de Trabajo No. 53*, 8–14. Lima: Instituto de Estudios Peruanos, 1994.

"¿De qué pedimos perdón?" *Hoy*, January 16, 1997.

Destler, I. M., and John Odell, assisted by Kimberly Ann Elliott. *Anti-Protection: Changing Forces in United States Trade Politics*. Washington, DC: Institute for International Economics, 1987.

Dixon, William J. "Democracy and the Peaceful Settlement of International Conflict." *American Political Science Review* 88, no. 1 (March 1994), 14–32.

Durán Barga, Jaime. "Actitud de los ecuatorianos frente al Perú: Estudio de opinión pública." In *Ecuador y Perú: Vecinos distantes*, 171–202. Quito: Corporación de Estudios para el Desarrollo, 1993.

"Economic Policies Criticized at Home and Abroad." *NotiSur Latin American Affairs* (Latin American Data Base) 7, no. 32 (September 5, 1997).

Ecuador. Constitutions 1979, 1993, 1998.

"Ecuador: Congress Votes to Oust President Abdalá Bucaram." *NotiSur* (February 7, 1997).

"Ecuador: Institutional Crisis Continues with Investigation of President Sixto Durán Ballén." *NotiSur* (November 10, 1995).

"Ecuador: Interim Government Calls Referendum as Supreme Court Orders Arrest of Former President." *NotiSur* (April 11, 1997).

"Ecuador-Peru: Quién ganó y quién perdió en la guerra empatada." *Hoy*, March 22 and April 3, 1995.

"Ecuador: President Sixto Durán Ballén Suffers Defeat in Referendum on Constitutional Reforms." *NotiSur* (December 1, 1995).

"Ecuador: Sixto Durán Ballén Turns Over Presidency to Successor Abdalá Bucaram." *NotiSur* (August 16, 1996).

"EE.UU. no ha intervenido en la crisis." *Hoy*, February 11, 1997.

Einaudi, Luigi R. "The Ecuador-Peru Peace Process." In Chester A. Crocker, Fen Osler Hampson, and Pamela Aall, eds., *Herding Cats: Multiparty Mediation in a Complex World*, 405–430. Washington, DC: U.S. Institute of Peace Press, 1999.

"El caballero de las dos torres." *Que Hacer* (Lima) 93 (January–February 1995), 18–19.

"En popularidad, Fujimori cede terreno a Andrade." *El Comercio* (Lima), March 21, 1999.

Escudé, Carlos. *Foreign Policy Theory in Menem's Argentina*. Gainesville: University Press of Florida, 1997.

Escudé, Carlos, and Andrés Fontana. "Argentina's Security Policies: Their Rationale and Regional Context." In Jorge I. Domínguez, ed., *International Security and Democracy: Latin America and the Caribbean in the Post–Cold War Era*, 51–79. Pittsburgh: University of Pittsburgh Press, 1998.

Espinosa, Carlos. "La negociación como terapia: Memoria, identidad y honor nacional en el proceso de paz Ecuador-Peru." In Adrián Bonilla, ed., *Ecuador-Perú: Horizontes de la negociación y el conflicto*, 111–138. Quito and Lima: FLACSO Sede Ecuador y DESCO Lima Perú, 1999.

———. "Memory, Identity, and Negotiation Dynamics." Seminar on Creating Peace: History, Negotiations, and the Peru-Ecuador Conflict, David Rockefeller Center of Latin American Studies (Harvard University, December 4, 1998).

Feaver, Peter D. *Armed Servants: Agency, Oversight, and Civil-Military Relations*. Cambridge, MA: Harvard University Press, 2003.

"Finance Minister on Cost of War, Reserve Reduction." *Voz de los Andes* in Spanish (Quito), March 2, 1995, in Foreign Broadcast Information Service—Latin America (FBIS-LAT-95–041, March 2, 1995, 54).

Fitch, J. Samuel. *The Military Coup d'Etat as a Political Process: Ecuador, 1948–1966*. Baltimore: Johns Hopkins University Press, 1977.

"Foreign Minister Concedes Protocol with Peru Valid." *Paris Agence France-Presse* in Spanish, January 26, 1995, in Foreign Broadcast Information Service—Latin America (FBIS-LAT-95–018, January 27, 1995, 30).

Foreign Relations of the United States (FRUS). Washington, DC: U.S. Government Printing Office, 1910.

Foreign Relations of the United States (FRUS). Washington, DC: U.S. Government Printing Office, 1937.

Forero, Juan. "Another Bump in a Rocky Road for Colombia and Venezuela." *New York Times*, November 30, 2000, A12.

Frankel, Jeffrey A. *Regional Trading Blocs in the World Economic System*. Washington, DC: Institute for International Economics, 1997.

Gallardo Ramón, Gen. José W. (Commander in Chief of the Army). "Comentario militar." In Hernán Alonso Altamirano Escobar, *El por que del ávido expansionismo del Perú*, 34–35. Quito: Instituto Geográfico Militar, 1991.

George, Alexander L., and Richard Smoke. *Deterrence in American Foreign Policy*. New York: Columbia University Press, 1974.

Glenn, Matt. "ICJ Begins Hearings in Argentina-Uruguay Paper Mill Dispute." *Jurist* (September 14, 2009). http://jurist.law.pitt.edu/paperchase/2009/09/icj-begins-hearings-in-argentina.php (accessed January 13, 2010).

Gobierno de Ecuador. Ministerio de Relaciones Exteriores. *Hacia la solución del problema territorial con el Perú: Libro blanco*. Quito: Ministerio de Relaciones Exteriores, 1992.

Gorman, Stephen M. "Geopolitics and Peruvian Foreign Policy." *Journal of Inter-American Economic Affairs* 36, no. 2 (Autumn 1982), 65–88.

Gourevitch, Peter. *Politics in Hard Times*. Ithaca: Cornell University Press, 1986.

"Government Issues Communique on Peruvian Border Incursions." *Voz de los Andes* in Spanish (Quito), January 25, 1995, in Foreign Broadcast Information Service—Latin America (FBIS-LAT-95-017, January 26, 1995, 54).

Graham, Carol. "Government and Politics." In Rex A. Hudson, ed., *Peru: A Country Study*, 205–263. 4th ed. Washington, DC: Federal Research Division, Library of Congress, 1993.

Griffin, Clifford E. *Power Relations and Conflict Neutralization in Latin America*. International Studies Working Paper. Stanford: Hoover Institution, March 1992.

Guarantor Countries of the Rio Protocol of 1942. "Communiqué by the Guarantor Countries of the Rio Protocol of 1942 to Ecuador and Peru." Brasilia, Brazil: Guarantor Countries, January 27, 1995.

"Guerra avisada no mata gente, las advertencias del Gral. Salinas Sedó." *Oiga* (Lima) 731 (February 20, 1995), 37–38.

"Habrá nuevas descalificaciones." *Hoy*, April 28, 1997.

Hermann, Charles F., Charles W. Kegley, and James N. Rosenau, eds. *New Directions in the Study of Foreign Policy*. Boston: Allyn & Unwin, 1987.

Hermann, Margaret G., Charles F. Hermann, and Joe D. Hagan. "How Decision Units Shape Foreign Policy Behavior." In Charles F. Hermann, et al., *New Directions in the Study of Foreign Policy*, 309–338. Boston: Unwin Hyman, 1989.

Hollist, W. Ladd, and Daniel L. Nelson. "Taking Stock of American Bonds: Approaches to Explaining Cooperation in the Western Hemisphere." *Mershon International Studies Review 42* (1998), 257–281.

Huth, Paul K. *Extended Deterrence and the Prevention of War*. New Haven: Yale University Press, 1988.

Informe Confidencial, January 6, 1995. Quito, Ecuador Archives.

Institute for Strategic Studies. *Military Balance, 1995–96*. London: Institute for Strategic Studies, 1997.

International Boundary Research Unit. http://www.dur.ac.uk/ibru/news/ (accessed June 23, 2011).

International Court of Justice. "Case concerning the Dispute regarding Navigational and Related Rights: Costa Rica v. Nicaragua." Judgment (International Court of Justice, July 13, 2009).

International Monetary Fund (IMF). *World Economic Outlook Database*. April 1999.

Isaacs, Anita. *Military Rule and Transition in Ecuador, 1972–92*. Pittsburgh: University of Pittsburgh Press, 1993.

"Island Dispute Sours Relations between El Salvador and Honduras." *Latin American Weekly Report*, WR-06-41, October 17, 2006, 16.

Jameson, Kenneth P. "Dollarization in Latin America: Wave of the Future or Flight to the Past?" *Journal of Economic Issues* 37, no. 3 (September, 2003), 643–664.

"Jueces unen a políticos." *Hoy*, October 4, 1996.

Korr, Charles P. *Cromwell and the New Model Foreign Policy: England's Policy toward France, 1649–1658*. Berkeley: University of California Press, 1975.

Krieg, William L. *Ecuadorean-Peruvian Rivalry in the Upper Amazon, Enlarged to Include the Paquisha Incident (1981)*. 2nd ed. Washington, DC: U.S. Department of State, 1986.

Laakso, Markku, and Rein Taagepera. "'Effective' Number of Parties: A Measure with Application to West Europe." *Comparative Political Studies* 12, no. 1 (1979), 3–27.

Lamas, Abraham. "Ecuador-Perú: Quién ganó y quién perdió en la Guerra Empatada." *InterPress Service* (March 10, 1995).

Lijphart, Arend. *Electoral Systems and Party Systems: A Study of Twenty-seven Democracies, 1945–1990*. New York: Oxford University Press, 1994.

Lindsay-Poland, John. "New Military Base in Colombia Would Spread Pentagon Reach throughout Latin America." *Americas Program*, Center for International Policy (May 28, 2009). http://www.cipamericas.org (accessed April 18, 2010).

"Los militares tras Rosalía." *Hoy*, February 11, 1997.

Loveman, Brian. *The Constitution of Tyranny: Regimes of Exception in Latin America*. 2nd ed. Pittsburgh: University of Pittsburgh Press, 1993.

———. *For la Patria: Politics and the Armed Forces in Latin America*. Wilmington, DE: Scholarly Resources, 1999.

Luna Tobar, Alfredo. "Vigencia e inejecutibilidad del Protocolo de 1942." *Revista de la Academia Diplomática* 1 (May 1996), 9–32.

Mainwaring, Scott, and Timothy R. Scully, eds. *Building Democratic Institutions: Party Systems in Latin America*. Stanford: Stanford University Press, 1995.

Mainwaring, Scott, and Matthew S. Shugart, eds. *Presidentialism and Political Parties in Latin America*. New York: Cambridge University Press, 1997.

Mansfield, Edward D., and Jack Snyder. "Democratization and the Danger of War." *International Security* 20, no. 1 (Summer 1995), 5–38.

———. "Democratization and War." *Foreign Affairs* 73, no. 3 (1995), 79–97.

Marcella, Gabriel. "Epilogue: The Peace of October 1998." In Gabriel Marcella and Richard Downes, eds., *Security Cooperation in the Western Hemisphere: Resolving the Ecuador-Peru Conflict*, 231–235. Coral Gables, FL: University of Miami, North-South Center Press, 1999.

———. *War and Peace in the Amazon: Strategic Implications for the United States and Latin America of the 1995 Ecuador-Peru War*. Carlisle, PA: Strategic Studies Institute, November 24, 1995.

———. *War without Borders: The Colombia-Ecuador Crisis of 2008*. Strategic Studies Institute. Carlisle, PA: U.S. Army War College, December 2008.

Mares, David R. "Boundary Disputes in the Western Hemisphere." *Pensamiento Propio* 14 (July–December 2001), 31–59.

———. "Civil-Military Relations, Democracy and the Regional Neighborhood." In David R. Mares, ed., *Civil-Military Relations: Building Democracy and Regional Peace in Latin America, Southern Asia and Central Europe*, 1–25. Boulder: Westview Press, 1998.

————. "Deterrence Bargaining in the Ecuador-Peru Enduring Rivalry: Designing Strategies around Military Weakness." *Security Studies* 6, no. 2 (Winter 1996–1997), 91–123.

————. "La disuasión y el conflicto Ecuador-Perú." Presentation at the Air Force War College, Quito, August 1995.

————. *Violent Peace: Militarized Interstate Bargaining in Latin America.* New York: Columbia University Press, 2001.

Martin, Lisa L. *Coercive Cooperation: Exploring Multilateral Economic Sanctions.* Princeton: Princeton University Press, 1992.

Masterson, Daniel M. *Militarism and Politics in Latin America: Peru from Sánchez Cerro to Sendero Luminoso.* New York: Greenwood Press, 1991.

Mauceri, Philip. *State under Siege: Development and Policy Making in Peru.* Boulder: Westview Press, 1996.

Maúrtua de Romaña, Oscar. "Hace cinco años estalló la paz." *Caretas* (Lima, October 23, 2003). http://info.upc.edu.pe/hemeroteca/tablas/actualidad/caretas/caretas1795.htm (accessed October 17, 2008).

McBride, George. "Ecuador-Peru Boundary Settlement." Report to the U.S. Secretary of State. Washington, DC: U.S. Department of State, 1949. Unpublished typescript.

McClintock, Cynthia. "Peru: Precarious Regimes, Authoritarian and Democratic." In Larry Diamond, Juan J. Linz, and Seymour Martin Lipset, eds. *Democracy in Developing Countries: Volume Four, Latin America*, 335–385. Boulder: Lynne Rienner Publishers, 1989.

————. "Presidents, Messiahs, and Constitutional Breakdowns in Peru." In Juan J. Linz and Arturo Valenzuela, eds., *The Failure of Presidential Democracy: The Case of Latin America*, 286–321. Baltimore: Johns Hopkins University Press, 1994.

McDermott, Jeremy. "U.S. Condemns Bolivia and Venezuela Ambassador Expulsions as 'Grave Error.'" *Telegraph*, September 12, 2008. http://www.telegraph.co.uk/news/worldnews/southamerica/bolivia/2825927/US-condemns-Bolivia-and-Venuzuela-ambassador-expulsions-as-grave-error.html (accessed April 18, 2010).

McDonnell, Patrick J. "Pardons Voided for Argentine 'Dirty War' Pair." *Los Angeles Times*, April 26, 2007, 3 (accessed via Lexis/Nexis, October 7, 2007).

Mearsheimer, John J. *Conventional Deterrence.* Ithaca: Cornell University Press, 1983.

Mecham, J. Lloyd. *The United States and Inter-American Security, 1889–1960.* Austin: University of Texas Press, 1962.

Mejía Acosta, Andrés. "¿Una democracia ingovernable?: Arreglos constitucionales, partidos políticos y elecciones en Ecuador, 1979–1996." B.A. thesis, Autonomous Technological Institute of Mexico, Mexico City, June, 1996.

Montezuma, Capt. Luis F. (Peru), and Lt. Gustavo Proaño (Ecuador). "Informe itinerario #1." Comisión Demarcadora de Límites, Subcomisión "El Cóndor," 2da Brigada, October 27, 1943.

Morey, Teodoro Hidalgo. *Las ganancias de Ecuador.* Lima: Producciones Gráficas "Borjas," 1997.

"Nicaraguan Missiles and Honduran Warplanes on Political Collision Course." *Noti-Cen*, February 8, 2007.

Niebieskikwiat, Natasha. "Hielos continentales: Reclamo de Chile por los mapas argentinos." *Clarin.com* (Argentina, August 29, 2006). http://www.clarin.com/diario/2006/08/29/elpais/p-01201.htm. (accessed March 30, 2008).

"No caminan las reformas." *Hoy*, April 25, 1997.

North, Douglass C. *Institutions, Institutional Change and Economic Performance.* Cambridge: Cambridge University Press, 1990.

NotiSur—Latin American Political Affairs (newsletter). Albuquerque, NM: Latin America Data Base, Latin American Institute, University of New Mexico.

Núñez, Ana. "Ecuador firma tratado de límites marítimos con nuestro país." *La República*, May 3, 2011.

Obando, Enrique. "Fujimori y las fuerzas armadas." In John Crabtree and Jim Thomas, eds. *El Perú de Fujimori*, 353–378. Lima: Instituto de Estudios Peruanos, 1999.

O'Donnell, Guillermo. "Delegative Democracy." *Journal of Democracy* 5, no. 1 (January 1994), 55–69.

Oneal, John R., and Bruce M. Russett. "The Classical Liberals Were Right: Democracy, Interdependence, and Conflict, 1950–1985." *International Studies Quarterly* 41, no. 2 (June 1997), 267–294.

"Opinión: Buenas intenciones." *Hoy*, January 15, 1997.

"Opinión: Ministerio de Etnias." *Hoy*, October 25, 1996.

"Opinión: Peregrinaje irresponsable." *Hoy*, February 20, 1997.

Owen, John M. "How Liberalism Produces Democratic Peace." *International Security* 19, no. 2 (Fall 1994), 87–125.

Palmer, David Scott. "El conflicto Ecuador-Perú: El papel de los garantes." In Adrián Bonilla, ed., *Ecuador-Perú: Horizontes de la negociación y el conflicto*, 31–60. Quito: FLACSO-Ecuador, 1999.

———. "National Security." In Rex A. Hudson, ed., *Peru: A Country Study*, 259–318. Washington, DC: Federal Research Division, Library of Congress, 1993.

———. "The Often Surprising Outcomes of Asymmetry in International Affairs: United States–Peru Relations in the 1990s." In Julio Carrion, ed., *The Fujimori Legacy: The Rise of Electoral Authoritarianism in Peru*, 227–241. University Park: Pennsylvania State University Press, 2006.

———. "Peru-Ecuador Border Conflict: Missed Opportunities, Misplaced Nationalism, and Multilateral Peacekeeping." *Journal of Interamerican Studies and World Affairs* 39, no. 3 (Fall 1997), 109–148.

———. "Peru's 1995 Elections." *LASA Forum*, 26, no. 2 (Summer 1995), 17–20.

———. "Peru's Persistent Problems." *Current History* 89, no. 543 (January 1990), 5–8ff.

———. *Revolution from Above: Military Government and Popular Participation in Peru, 1968–1972.* Ithaca: Cornell University Latin American Studies Program, 1973.

———. "Revolution in the Name of Mao: Rebellion and Response in Peru." In Robert Art and Louise Richardson, eds., *Democracy and Counterterrorism: Lessons from the Past*, 195–220. Washington, DC: U.S. Institute of Peace Press, 2007.

————. "The Search for Conflict Resolution: The Guarantors and the Peace Process in the Ecuador-Peru Dispute." In Gabriel Marcella and Richard Downes, eds., *Security Cooperation in the Western Hemisphere: Resolving the Ecuador-Peru Conflict*, 21–44. Miami: North-South Center Press, 1999.

Paul, T. V. *Asymmetric Conflicts: War Initiation by Weaker Powers*. Cambridge: Cambridge University Press, 1994.

Pear, Nicky, and Alexandra Reed. "Dredging Up an Old Issue: An Analysis of the Long-standing Dispute between Costa Rica and Nicaragua over the San Juan River." *Council on Hemispheric Affairs*, January 24, 2011. http://www.coha.org/dredging-up-an-old-issue-an-analysis-of-the-long-standing-dispute-between-costa-rica-and-nicaragua-over-the-san-juan-river-2/ (accessed June 26, 2011).

Perkins, Whitney T. *Constraint of Empire*. West Haven, CT: Greenwood Press, 1981.

"Peru: Foreign Minister Resigns in Midst of Negotiations with Ecuador." *NotiSur—Latin American Affairs* 8, no. 37 (October 9, 1998).

"Peru Takes Possession of Chilean Port Terminal: Access to Arica Promised since 1929." *Santiago Times*, February 15, 2000.

"Peru Was Preparing for War with Chile, Reveals President Fujimori." *CHIP News*, March 3, 1995.

Pike, Frederick B. *The United States and the Andean Republics: Peru, Bolivia, and Ecuador*. Cambridge, MA: Harvard University Press, 1977.

Pion-Berlin, David. "From Confrontation to Cooperation: Democratic Governance and Argentine Foreign Relations." In David R. Mares, ed., *Civil-Military Relations in Latin America: New Analytical Perspectives*, 79–100. Chapel Hill: University of North Carolina Press, 2001.

Pion-Berlin, David, and Harold A. Trinkunas. "Attention Deficits: Why Politicians Ignore Defense Policy in Latin America." *Latin American Research Review* 42, no. 3 (2007), 76–100.

Pons Muzzo, Gustavo. *Estudio histórico sobre el Protocolo de Rio de Janeiro*. Lima: n.p., 1994.

Prieto Celi, Federico. *El golpe*. Lima: B&C Editores, 1992.

"Pulp-Mill Verdict Satisfies Honour on Both Sides of River Uruguay." *Latin American Weekly Report* WR-10–16 (April 22, 2010), 1.

Ratner, Steven R. "Drawing a Better Line: *Uti Possidetis* and the Borders of New States." *American Journal of International Law* 90, no. 4 (October 1996), 590–624.

"Reforma contra Bucaram." *Hoy*, February 19, 1997.

Reid, Angus. "Colombia, Ecuador Assess Cross-Border Incursion." *Global Monitor, Polls & Research* (poll conducted by Centro Nacional de Consultoría and Cedatos-Gallup released by CM, March 9, 2008). http://www.angus-reid.com/polls/view/30098/colombia_ecuador_assess_cross_border_incursion (accessed April 2, 2008).

República del Ecuador. "Plan Binacional Capítulo Ecuador." 2008. http://www.planbinacional.gov.ec/ (accessed October 15, 2008).

República del Ecuador. Ministerio de Relaciones Exteriores. *Ecuador: Impases subsistentes*. Quito: Ministerio de Relaciones Exteriores, 1996.

————. *Hacia la solución*. Quito: Ministerio de Relaciones Exteriores, 1992.

————. *Misión en Washington.* Quito: International Court of Justice, 1981.

República del Perú. Ministerio de Relaciones Exteriores. *Ayuda memoria: Desacuerdos sobre demarcación de la frontera.* Lima: Ministerio de Relaciones Exteriores, n.d.

————. *Frontera peruano-ecuatoriana: El laudo arbitral de Braz Dias de Aguiar— Reportorio documental.* Edición Especial, 40° Aniversario de la Academia Diplomática del Perú. Lima: Academia Diplomática del Perú, 1996.

Rodríguez Elizondo, José. *Chile-Perú: El siglo que vivimos en peligro.* Santiago: La Tercera-Mondadori, 2004.

Rojas Aravena, Francisco. "América Latina: Alternativas y mecanismos de prevención en situaciones vinculadas a la soberanía territorial." Paper prepared for the Carter Center, Atlanta, GA. Published in *Paz y seguridad en las Américas,* 5–8. Santiago: FLACSO-Chile, 1997.

————. "Transition and Civil-Military Relations in Chile." In Jorge I. Domínguez, ed., *International Security & Democracy: Latin America and the Caribbean in the Post-Cold War Era,* 80–101. Pittsburgh: University of Pittsburgh Press, 1998.

Romero, Simon. "Increased U.S. Military Presence in Colombia Could Pose Problems with Neighbors." *New York Times* (July 22, 2009). http://www.nytimes.com/2009/07/23/world/americas/23colombia.html (accessed April 18, 2010).

————. "Venezuela Spending on Arms Soars to World's Top Ranks." *New York Times* (February 25, 2007). http://www.nytimes.com/2007/02/25/world/americas/25venez.html?pagewanted=1&ei=5088&en=25be50a9a3dfea93&ex=1330059600 (accessed May 21, 2007).

Rosales Ramos, Francisco. "Dilema." *Hoy,* February 24, 1977.

Rosenberg, Mark, et al. *Honduras: Pieza clave de la política de Estados Unidos en Centro América.* Tegucigalpa: Centro de Documentación de Honduras (CEDOH), 1990.

Rouquié, Alain. *The Military and the State in Latin America.* Translated by Paul E. Sigmund. Berkeley: University of California Press, 1987.

Saad Herrería, Pedro. *La caída de Abdalá.* Quito: El Conejo, 1997.

Saba, Raúl P. *Political Development and Democracy in Peru: Continuity in Change and Crisis.* Boulder: Westview, 1987.

Sánchez, Alex. "Costa Rica: An Army-less Nation in a Problem-Prone Region." *Council on Hemispheric Affairs,* June 2, 2011. http://www.coha.org/costa-rica-an-army-less-nation-in-a-problem-prone-region/ (accessed June 26, 2011).

Scheggia Flores, Carlos E. *Origen del pueblo ecuatoriano y sus infundadas pretensiones amazónicas.* Lima: Talleres de Línea, 1992.

Schumacher, Edward. "Behind Ecuador War, Long-Smoldering Resentment." *New York Times,* February 10, 1981, A2.

Sereseres, Caesar. "The Interplay of Internal War and Democratization in Guatemala since 1982." In David R. Mares, ed., *Civil-Military Relations: Building Democracy and Regional Peace in Latin America, Southern Asia and Central Europe,* 206–222. Boulder: Westview Press, 1998.

Shugart, Matthew Soberg, and John M. Carey. *Presidents and Assemblies: Constitutional Design and Electoral Dynamics.* Cambridge: Cambridge University Press, 1992.

Simmons, Beth A. "Trade and Territorial Conflict in Latin America: International Borders as Institutions." In Miles Kahler and Barbara F. Walter, eds., *Territoriality and Conflict in an Era of Globalization*, 251–287. Cambridge: Cambridge University Press, 2006.

Snyder, Jack. *Myths of Empire*. Ithaca: Cornell University Press, 1991.

Spindler, Frank MacDonald. *Nineteenth Century Ecuador: A Historical Introduction*. Fairfax, VA: George Mason University Press, 1987.

Steinmo, Sven, Kathleen Thelen, and Frank Longstreth, eds. *Structuring Politics: Historical Institutionalism in Comparative Analysis*. Cambridge: Cambridge University Press, 1992.

Stepan, Alfred. *The Military in Politics: Changing Patterns in Brazil*. Princeton: Princeton University Press, 1971.

St. John, Ronald Bruce. "The Boundary between Ecuador and Peru." *Boundary & Territorial Briefing* 1, no. 4 (Durham, UK: International Boundaries Research Unit, University of Durham, 1994), 1–24.

———. "Conflict in the Cordillera del Cóndor: The Ecuador-Peru Dispute." *Boundary and Security Bulletin* 4, no. 1 (Spring 1996), 78–85.

———. *The Foreign Policy of Peru*. Boulder: Lynne Rienner Publishers, 1992.

Stokes, Susan. "Peru: The Rupture of Democratic Rule." In Jorge I. Domínguez and Abraham F. Lowenthal, eds., *Constructing Democratic Governance: South America in the 1990s*, 58–71. Baltimore: Johns Hopkins University Press, 1996.

Taagepera, Rein, and Matthew Sobert Shugart. *Seats and Votes: The Effects and Determinants of Electoral Systems*. New Haven: Yale University Press, 1989.

Tilly, Charles. "War Making and State Making as Organized Crime." In Peter B. Evans, Dietrich Rueschemeyer, and Theda Skocpol, eds., *Bringing the State Back In*, 169–187. Cambridge: Cambridge University Press, 1985.

"Tiwinza: Descansa en paz." *Oiga* 731 (Lima, February 20, 1995), 13–19.

Tobar Donoso, Julio, and Alfredo Luna Tobar. *Derecho territorial ecuatoriana*. 4th ed. Quito: Ministry of Foreign Affairs, 1994.

Tuesta Soldevilla, Fernando. *Perú político en cifras*. 2nd ed. Lima: Fundación Friedrich Ebert, 1994.

U.S. Arms Control and Disarmament Agency (ACDA). *World Military Expenditures and Arms Transfers, 1985*. Washington, DC: U.S. Government Printing Office, 1986.

———. *World Military Expenditures and Arms Transfers, 1993–94*. Washington, DC: U.S. Government Printing Office, 1995.

U.S. Department of Defense. *Maritime Claims Reference Manual*. Washington, DC: Department of Defense, 2005.

———. *Maritime Claims Reference Manual, June 2008*. http://www.dtic.mil/whs/directives/corres/html/20051m.htm (accessed April 12, 2010).

———. *United States Security Strategy for the Americas*. Washington, DC: Pentagon, Office of International Security Affairs, 1995.

U.S. Embassy of Peru. *The 1995 Peruvian-Ecuadorean Border Conflict*. Washington, DC: Embassy of Peru, March 1995.

Vargas Pazzos, Lt. Gen. Frank. *Tiwintza: Toda la verdad*. Quito: Color Gráfica, 1995.

Villanueva, Víctor. *100 años del ejército peruano: Frustraciones y cambios*. Lima: Editorial Juan Mejía Baca, 1971.

———. *El CAEM y la revolución de la fuerza armada*. Lima: Instituto de Estudios Peruanos, 1973.

Vlahos, Alexia. "Bolivia Demands Access to Pacific Ocean: Arica Tunnel." *Reuters* (January 4, 2010). http://www.reuters.com/article/2010/01/04/us-chile-tunnel-idUSTRE6032VL20100104 (accessed January 13, 2010).

Weidner, Col. Glenn R. "Operation Safe Border: The Ecuador-Peru Crisis." *Joint Forces Quarterly* (Spring 1996), 52–58.

———. "Peacekeeping in the Upper Cenepa Valley: A Regional Response to Crisis." Paper presented at the Conference on Multilateral Approaches to Peacemaking and Democratization in the Hemisphere (North-South Center, University of Miami, April 11–13, 1996). A portion of this paper was published as "Operation Safe Border: The Ecuador-Peru Crisis." *Joint Forces Quarterly* (Spring 1996), 52–58.

———. "Peacekeeping in the Upper Cenepa Valley: A Regional Response to Crisis." In Gabriel Marcella and Richard Downs (eds.), *Security Cooperation in the Western Hemisphere: Resolving the Ecuador-Peru Conflict*, 45–66. Miami: North-South Center Press, University of Miami, 1999.

Wise, Carol. "Against the Odds: The Paradoxes of Peru's Economic Recovery in the 1990s." In Julio F. Carrión, ed., *The Fujimori Legacy: The Rise of Electoral Authoritarianism in Peru*, 201–226. University Park: Pennsylvania State University Press, 2006.

Wood, Bryce. *The United States and Latin American Wars, 1932–42*. New York: Columbia University Press, 1966.

Zibechi, Raúl. "Is Brazil Creating Its Own 'Backyard'?" *Zibechi Report* 12 (Americas Program, February 3, 2009).

Interviews

Alcorta Silva Santisteban, Edith (Peruvian Office of the Binational Development Plan). Palmer interview, Lima, Peru, August 15, 2008.

Azabache, Cesar (Defensoría del Pueblo). Mares interview, Lima, Peru, April 5, 1999.

Bonilla, Adrián. Mares interview, Guadalajara, Mexico, April 18, 1997.

Chamochumbi, Gen. Carlos (Peru, retired; commander of the Peruvian army division in the border city of Tumbes in 1991–1992). Palmer interview, Lima, Peru, August 14, 2008.

Confidential interview with a former high-ranking Ecuadorean diplomat. Mares interview, Quito, Ecuador, August 1995.

Confidential interview with a former high-ranking Peruvian diplomat. Mares interview, Lima, Peru, March 26, 1999.

Confidential interview with a Latin American diplomat. Palmer interview, Santiago, Chile, August 23, 1997.

Confidential interview with an Ecuadorean diplomat. Mares interview, Quito, Ecuador, August 1995.

Confidential interview with a Peruvian analyst. Mares interview, Quito, Ecuador, August 1995.

Confidential interview with Peruvian career diplomats. Letter to Palmer, August 26, 1996.

Confidential interview with U.S. military analysts. Mares interview, Washington, DC, September 1995.

Confidential interviews with Ecuadorean military officers. Mares interviews, Quito, Ecuador, August 1995.

Confidential interviews with two U.S. government analysts. Palmer interviews, Washington, DC, September 1997.

Costa, Dr. Ferrero. Palmer interview, Peruvian Embassy, Washington, DC, September 9, 1997.

Durán, Jaime (director of *Informe Confidencial*). Mares interview, Quito, Ecuador, July 15, 1997.

Einaudi, Ambassador Luigi (U.S. guarantor representative). Palmer interview, Washington, DC, January 29, 1997.

Gallardo, Gen. José W. (minister of defense). Mares interview, Quito, Ecuador, August 1995.

Hernández, Col. Luis B. (personal secretary to the minister of defense of Ecuador and commander of the Tiwintza defense during the war). Mares interview, Quito, Ecuador, August 14, 1995.

Hollis, Caryn (U.S. Defense Intelligence Agency analyst). Palmer interview, Washington, DC, June 23, 1992.

Huerta, Luis (Comisión Andina de Juristas). Mares interview, Lima, Peru, March 25, 1999.

Krieg, William L. Palmer interview, Sarasota, FL, September 21, 1996.

Luna Tobar, Ambassador Alfredo. Mares interview, Quito, Ecuador, August 17, 1995.

Mahuad, Jamil (former president of Ecuador). Palmer interview, Cambridge, MA, July 11, 2008.

Moncayo, Francisco "Paco" (army chief general, Ecuador). Mares interview, Quito, Ecuador, August 16, 1995.

Ponce, Javier (deputy chief of the Ecuadorean Mission to the United Nations). Palmer interview, New York City, February 13, 1995.

Proaño, Dr. Luis Heladio (political advisor, Ministry of Defense, Ecuador). Mares interview, Quito, Ecuador, August 14, 1995.

Rojas Escalante, Franklin (Peruvian office of the Binational Development Plan). Palmer interview, Lima, Peru, August 15, 2008.

Sicade, Lynn (U.S. diplomat). Palmer interview, Washington, DC, April 2, 1996.

Terán, Edgar (Ecuadorean ambassador to the United States). Palmer interview, Washington, DC, April 3, 1996.

Vargas Pazzos, Gen. Frank (Ecuador). Mares interview, Quito, Ecuador, August 16, 1995.

Villegas, Col. Juan (Ecuador: member of army chief general Paco Moncayo's personal staff as a major during the first months of the 1995 war). Palmer interviews, Quito, Ecuador, February 13 and 28, 2008.

Watson, Ambassador Alexander (assistant secretary of state for American republics affairs, July 1993 to March 1996). Palmer interview, Arlington, VA, October 24, 2001.

Williams Zapata, Gen. José D. (Peru, retired; 1995 commander of the special forces brigade in the Cenepa). Palmer interview, Washington, DC, December 4, 2008.

INDEX

The letter *t* following a page number denotes a table.
The letter *f* denotes a figure.